Editor in Chief of Book Series: Wang Guoliang
Series of Theory and Method on Disaster Prevention and Mitigation,
Post-disaster Reconstruction and Poverty-reduction and Development

Disaster Response and Rural Development

—— International Symposium on Theories and Practices of
Disaster Risk Management and Poverty Reduction

Huang Chengwei, Lu Hanwen Eds.

Huazhong Normal University Press

图书在版编目(CIP)数据

灾害应对与农村发展:"灾害风险管理与减贫的理论及实践"国际研讨会论文集＝Disaster Response and Rural Development—International Symposium on Theories and Practices of Disaster Risk Management and Poverty Reduction:英文/黄承伟,陆汉文主编
—武汉:华中师范大学出版社,2013.5
(防灾减灾、灾后重建与扶贫开发理论方法研究丛书)
ISBN 978-7-5622-6074-5

Ⅰ.①灾… Ⅱ.①黄… ②陆… Ⅲ.①自然灾害—风险管理—中国—国际学术会议—文集—英文 Ⅳ.①X432-53 ②F124.7-53

中国版本图书馆 CIP 数据核字(2013)第 114508 号

Authors: Huang Chengwei & Lu Hanwen
Executive Editor: Zhou Kongqiang   Liu Xiaojia
Chief Proofreader: Wang Sheng
Cover Designer: Gan Ying
Editorial Office Tel: +8602767867364
Published by Huazhong Normal University Press
Address: No. 152 Luoyu Road, Wuhan, Hubei Province, China
Tel: +8602767863040 (Publishing Office) +8602767861321(Mail-order Office)
Fax: +8602767863291
Website: www.ccnupress.com   Email: hscbs@public.wh.hb.cn
Printer: Huazhong University of Science and Technology Printing Co. Ltd
Print Supervisor: Zhang Guangqiong
Print Sheet: 19.5
Format: 710×1000, 16mo
Edition: First Edition, June 2013   Impression: First Printing, June 2013
Price: 45.00 RMB

Welcome to search and buy books on our website.
Welcome to report pirated books, and please telephone informants' hot-line 027-67861321.

Series of Studies on Disaster Prevention and Mitigation, Post-disaster Reconstruction and Poverty-reduction and Development Editorial Board

Director: Fan Xiaojian
Vice Director: Wang Guoliang   Zheng Wenkai
Members (In order of the surname strokes):
　　　　　Wang Guoliang   Si Shujie   Li Chunguang   Fan Xiaojian
　　　　　Zheng Wenkai   Hong Tianyun   Hai Bo   Xia Gengsheng
　　　　　Jiang Xiaohua

Editor in Chief of the Book Series: Wang Guoliang

CONTENTS

A Study on the Theories and Practice of Disaster Risk Management and Poverty Reduction: A New Issue ·············· Huang Chengwei Li Haijin (1)

Strategic Thinking on the Combination of the Disaster Risk Prevention with Poverty Reduction in Poor Villages in Special Areas—A Case Study of Plateaus Areas Such As Yushu, Qinghai Province
·· Zhang Qi   Wang Hao (33)

Passion, Ideal and Reality—the Relationship between One Non-governmental Organization and Rural Communities during the Process of Post-earthquake Reconstructions and Its Significance
··· Lu Hanwen   Yue Yaopeng (56)

Analysis of the Impact of Household Endowment on Responding Capacity of Peasant Households to Disaster Risk—Based on an Investigation of 39 Poor Ethnic Counties in Southwestern China
··· Zhuang Tianhui   Zhang Haixia (86)

Strengthen Capacity of Poverty Reduction by Village Public Products Supply
—Taking Jiezhu Village, Xi'eluo Town of Yajiang County in Ganzi Tibetan Autonomous Prefecture of Sichuan Province as an Example
································· Li Xueping   Longming Azhen (112)

Mechanism Innovation and Enlightenment to the Combination of Post-earthquake Reconstruction with Development-oriented Poverty Reduction for Poverty-stricken Villages in the Earthquake-stricken Wenchuan of Sichuan Province ················· Xiang Xinghua   Qin Zhimin (140)

Sustainable Utilization of Natural Resources and the Environmental Risk

Management Strategies of Poverty Alleviation and Development—A Case Study of Sustainable Use of Wild Chinese Herbal Medicine in Daping Village, Pingwu County, Sichuan Province ·········· Deng Weijie (163)

A Cost-benefit Analysis of Disaster Risk Reduction
··· Christoph I. Lang (178)

DFID's Work & Experience on Disaster Risk Management and Poverty Reduction ··· DFID (188)

Disaster Risk Management and Poverty Reduction: International Experiences
··· Sanny R. Jegillos (191)

Basis and Prospect of Pilot Poor Villages' Sustainable Livelihoods in Post-reconstruction Period—An Analysis Based on the Second Annual Comprehensive Assessment Data ························ Cai Zhihai (199)

Significance and Benefit Evaluation of the Project of Returning Farmland to Forest and Grassland in Western Poverty-stricken Areas—Taking the Longnan city in Gansu Province as an Example
································· Wang Jianbing Tian Qing (219)

Japan's Disaster Management and Its Implications for China
························ Cheng Guangshuai Liang Hui (227)

A Study on the Influence of Natural Disaster on Clustering Contiguous Special Poverty-Stricken Communities—Taking the Wuling Mountainous Area as an Example ··· Zhang Dawei (245)

Paradigm Shift of Disaster Risk Management: Looking upon Disaster Mitigation from the Perspective of Poverty Reduction ·············· Lv Fang (267)

The Unconventional Action on Impoverished Villagers' Housing Reconstruction in Earthquake-stricken Areas—Taking Makou Village in Sichuan Province as an Example ································· Chen Wenchao (284)

Postscript ··· (306)

A Study on the Theories and Practice of Disaster Risk Management and Poverty Reduction: A New Issue

Huang Chengwei　Li Haijin[*]

Abstract: In the new period of frequent natural disasters, the practical work and theoretical propositions of disaster risk management and poverty alleviation appear as a highly innovative and leading academic thesis. The article, based on existing resources and social practice, constructs a set of analysis framework on disaster risk management and poverty reduction, clarifies their logical relation, makes preliminary explanation for the current disaster risk management, compares the disaster risk management and poverty alleviation at home and abroad from the international perspective to use the successful experience for reference.

Key words: Disaster Risk; Disaster Risk Management; Poverty Reduction; Poverty Vulnerability

Currently, with climate change, the global is gradually entering into a period when natural disasters happens frequently, and

---

[*] Huang Chengwei, Deputy Director of International Poverty Reduction Center in China (IPRCC), Doctor, Researcher; Li Haijin, Associate Professor of the Institute of Chinese Rural in Central China Normal University, Visiting Researcher of International Poverty Reduction Center in China.

fundamental changes also occur on the quality, types, characteristics of the disasters, the economic and social harm are also more notable. What is most badly in need of attention is that disasters not only perform a real form and tangible fact, but also more and more highlight a potential risk state, resulting in new appearance of the society with unexpected and endless high-risk. Meanwhile, in terms of the practical situation, disasters and the risk mainly exist in key areas of poor areas and poor people. In the above background, the practical work and theoretical propositions of disaster risk management and poverty reduction appear as a highly innovative and leading academic thesis. So the author attempts to base on existing resources and social practice, constructs a set of analysis framework on disaster risk management and poverty reduction, clarifies their logical relation, makes preliminary explanation for the current disaster risk management, compares the disaster risk management and poverty at home and abroad from the international perspective so that the successful experience can be sifted out.

1. Research Background and Issues

This study is basically based on the following background: first, the disasters and its risk status; second, the consequence of disasters; third, the relationship between disaster and poverty; Fourth the relationship between disaster risk management and poverty reduction.

Firstly, the global is entering into a period when natural disasters happen frequently. When coming to twenty-first century, particularly in recent years, natural disasters frequently occurred around the world. A variety of signs indicate that the world has entered the period of frequent occurrence of natural disasters. The potential environment risk which human society faces is considerably serious. The continuously emergence of new types of disasters, disaster losses, and all these severely threaten every country and region in the world. Especially the nearly unpredictable disasters triggered by climate changes make the original fragile surviving situation more severe. Disaster problem has been a significant global issue which attracts extensive attention of worldwide in current society.

Secondly, disasters lead to tremendous and far-reaching economic and social losses. The frequent occurrence of natural disasters not only seriously threatens people's lives and health, damages people's production and living order and destructs existing resources and environment, but also affects the social development and progress. What is worse, it even leads to the destruction of civilization. Because of the complexity and diversity of topography, China has always been a country with frequent occurrence of natural disasters, which is characterized by, high frequency, wide distribution and great losses. ① Particularly, earthquake disasters and corresponding secondary hazards which features sudden, extensive and strong destructiveness in the latest few years gave rise to substantial property damage and casualties. It brought immense and profound loss to economic and social development and especially exert far-reaching influence on sustainable development and property relief efforts of poverty-stricken areas.

Thirdly, disaster has a high degree of correlation with poverty. The occurrence causes of several major natural disasters and subsequent coping strategies adopted shows disaster and poverty have a high degree of correlation at two levels of practice and policy framework. First of all, the enormous destructiveness reduced disaster-stricken areas and victims into poverty status or the edge of poverty. It increased the poverty level of victims who are dwelling in poor areas. The so-called disasters cause poverty and returning to poverty means that disaster is becoming an important new factor which contributes to poverty. The recovery of economy and society in developed areas is relatively easier in the post-disaster construction. However, it is extremely difficult to go back to the previous state for poverty-stricken areas. And it undermined the existing poverty reduction achievement to a large extent and decelerated the process of poverty reduction. Secondly, there is a high degree of overlap in the distribution of poverty and disaster striken

---

① Deng Tuo. *History of Famine Relief in China*. Beijing: Beijing Press, 1998.

areas. Areas with many geological hazards are always impoverished regions with harsh natural environment, which further aggravates the vulnerability and extent of the damage in poverty-stricken areas and victims and reduce its disaster response capacity against disasters. Last but not the least, natural disaster not only has natural properties but also possesses distinct social attributes in the new historical background. It is the result that natural factors and human factors contributes should both be to blame.

Fourthly, disaster risk management and poverty reduction are inseparable and mutually reinforcing. It challenges the poverty reduction work and also raises the awareness of new understanding of poverty reduction from the perspective of disaster risk management. Scientific and effective disaster risk management can be conducive to reducing the cost of poverty alleviation, improving the effects of poverty reduction and consolidating the achievement. And highly effective poverty reduction work could in turn contributes to disaster prevention and mitigation. Judging from the current research situation at home and abroad, starting research on disaster risk management and poverty reduction can be conducive to establishing new types of strategies and policy-making mechanism in reaction to disasters constant innovation of the institutions and mechanisms for coping with disasters and constructs disaster risk management theories and measures with stronger integration and higher validity. Furthermore, It provides pioneering theoretical vision and techniques. Therefore, re-examining the disaster issues in China's rural poverty-stricken areas and investigating the disaster risk coping mechanisms which are suitable for rural impoverished district make disaster risk management truly integrated into a new phase of the Poverty Reduction Strategy. It will be a fundamentally effective route for poverty alleviation in poor areas with frequent disasters.

Study on the theories and practice of disaster risk management and poverty reduction is a brand-new research field with great potential. It not only has profound theoretical value, but also has important practical meaning and policy implications, especially has great significance for international exchange.

Firstly, as a new research field, study on the theories and practice of disaster risk management and poverty reduction reflects a strong theoretical necessity and value. Literature review shows that individual research achievements related to disaster risk management and poverty reduction are considerably copious, however, comprehensive study which links disaster risk management with poverty reduction is rarely carried out. It is necessary to carry out theoretical combination and explanation for this new proposition and conduct theoretical improvement combining current mitigation and poverty reduction, which will make contribution to theoretical innovation of global disaster mitigation and poverty reduction. At the same time, theories are bound to regard practice and policy as direction and target. The study can offer fundamental theoretical support to practice and policy for disaster risk management and poverty reduction. In the context of risk society, the endorsement of new development-oriented theories are particularly required by poverty-stricken areas such as multi-factor superimposed districts with high disaster risk, environment vulnerability, wide range of poverty and deep level of poverty.

Secondly, study on the theories and practice research on disaster risk management and poverty reduction provides a forward-looking, targeted and effective coping strategies and route for disaster risk management and poverty reduction work. Natural disasters mostly occurs in the impoverished areas which has a high degree of superposition and accumulation of ecological vulnerability, economic vulnerability and social vulnerability. This greatly aggravates the depth and breadth of poverty and also poses more severe challenges to poverty-reduction work. The work which will focus on development-oriented poverty reduction over the next decade will target the regions with special difficulties that lie in vast and contiguous stretches of frequent disasters and high vulnerability. We ought to transform our working mechanism and combine disaster risk management with development-oriented poverty reduction if the effectiveness of poverty alleviation in these districts needs to be improved. We must accomplish the target of effective poverty reduction and achieve win-win situation when responding effectively and orderly to management of disasters risk. Simultaneously, the idea and practice of disaster risk management is consistent with the inclusive growth and transformation of economic development patterns proposed by central authority. And it also provides better paths and mechanisms for the latter' practical implementation.

In addition, studies on the theories and practice research on disaster risk management and poverty reduction lay a solid foundation for formulation and adjustment of poverty reduction policy. On the one hand, the Outline for Development-oriented Poverty Reduction for China's Rural Areas necessarily incorporates disaster prevention and mitigation, disaster risk management and development-oriented alleviation in the process of drafting. It also focuses on the regions with special difficulties in ecological vulnerability, disaster risk and high degree of poverty especially in vast and contiguous stretching areas. And it naturally requires accurate, precise and appropriate strategic location and technological support to this policy program in macro-level theory. On the other hand, the poverty alleviation policy over the next decade is likely to be adjusted according to the changing reality. In the process of policy regulation, some forward-looking issues or innovative areas such as disaster risk management are generally difficult to reach a consensus among all relevant government departments and working agencies in a relatively short period. It requires corresponding support and foundation for policy communication at the research level.

Finally, disaster risk management, theories and practical research of poverty reduction build a favorable platform for international exchange and cooperation. Our study will face the international society, get lessons from international experience to achieve the internationalization of dementia experience and start international exchange and cooperation. We'll acquire and take advantage of international academic resources to enhance research standards for disaster risk management and poverty reduction theories in China. We can learn from international experience, promote the practice which combines China's disaster risk management with poverty reduction improving its operational performance. We can also refine available practice and experience for international communication in terms of combining disaster risk management with poverty reduction of China. It develops international filed expedition, learning and training in disaster risk management and poverty reduction. The study is also a demand of fulfilling its international obligations, establishing a major power image and heightening the content of soft power from the main point view of China. International community including the majority of developing countries comes to a highly consistent evaluation for China's achievements of poverty reduction. They have a strong desire to learn lessons from China's poverty reduction experience. It is of great practical and strategic significance in coordinating the country's diplomatic strategies in poverty reduction areas and increasing the intensity of output of poverty reduction and development experience.

2. Core Concepts

2.1 Poverty, poverty reduction

Poverty and poverty reduction are the most basic and core concepts as well as the most controversial concepts in the research of poverty and anti-poverty. It experiences notable changes namely, from income poverty that pays attention to single, surface income and consumption factors, to ability poverty that highlights social factors, such as health, education and so on, then to rights poverty or human poverty that emphasizes political, psychological, cultural factors such as social vulnerability, social exclusion, deprivation of sex and political rights. Correspondingly, cognition and understanding of poverty change from absolute poverty to relative poverty, from short-term poverty (or temporary poverty) to long-term poverty (or persistent poverty). [1]

---

[1] Guo Xibao & Luo Zhi. *Discussing the Evolution of the Concept of Poverty*. Jiangxi Social Science, 2005; Gao Yunhong & Zhang Jianhua. *Evolution of the concept of poverty*. Reformation, 2006.

Under this background, the theory circle and the practice circle are more and more likely to analyse poverty and its related topics from multidimensional perspectives and multiple levels, developing concepts and analytical paradigm of multi-dimensional poverty. Knowledge poverty, information poverty and even ecological poverty are also gradually entering into people's view.① Diversification and multidimensionality of poverty have decisive effects on poverty reduction strategies and choice of policy instruments.

Poverty reduction, cognizes poverty from policies and actions. In brief, poverty reduction refers to reducing poverty or alleviating poverty, namely reducing or alleviating poverty in terms of its coverage, population, degree and depth with the help of policy instruments. Its main evaluation indexes are poverty population and poverty incidence.

2.2 Disaster, disaster risk, disaster risk management

Generally speaking, disaster is a generic term of incidences that bring human and human's survival environment destructive impacts. Natural disasters have obvious, also marked, far-reaching effects on human's production and normal life as well as economic and social development. The strong desire to scientifically understand natural disasters, effectively prevent and respond to natural disasters so that the development of population and environment can be sustainable and positively interact with each other, makes social science circle's theoretical propositions more and more crucial. In modern society, the disaster increasingly appears as a potential risk state, disasters occur unnecessarily but constantly being in an unpredictable and unexpected

---

① Shang Weiping & Yao Zhimou. *Multidimensional Poverty Measurement Method Research*. Finance and Economy Research, 2005; Chen Lizhong. *Multidimensional Poverty Estimate and Decomposition of China in Transition Period*. Economic Review, 2008; Wang Xiaolin & Sabina Alkire. *Multidimensional Poverty Survey: Meanings of Estimate and Policy*, Rural Economy in China, 2009; Hu Angang & Tong Xuguang. *China's Poverty Reduction Theory and Practice—in Qinghai's Perpective*. Proceedings of Tsinghua University(Philosophy and Social Science Edition), 2010; Hu Angang & Li Chunbo. *New Poverty in New Century: Knowledge Poverty*. China's Social Science, 2001.

state with extreme instability. Definition of disaster risk consists of key elements, namely occurrence rate, triggering agent, invulnerability, and can be summarized as three levels, namely define disaster risk as the loss of certain rate conditions from disaster's own perspective, define disaster risk as the rate of triggering agent's appearance from triggering agent's perspective, put emphases on invulnerability of human society's economy and enlarged or step-down effect for disasters by human's activities then define disaster risk as combination of triggering agent and invulnerability. According to the above academic context, we can define disaster risk as "possibility of loss and damage due to combined actions of invulnerability of various triggering agents and human system"[1].

Putting forward the concept of disaster risk management is mainly aimed at rethinking recovery reconstruction after a heavy disaster, prevention's emergency management before a trivial disaster and crisis management which stresses result and despises process, trying to establish a set of comprehensive, sustainable, multiagent, crisis-cross management system. We define disaster risk management as human recognition, estimation, evaluation of potential various disaster risk, and on that basis synthetically utilize law, administration, economy, technology, engineering, through the integration of the organization and social cooperation, through the whole process of disaster management, to enhance ability of government and society's disaster risk management as well as disaster prevention and reduction, effectively preventing, responding, reducing all kinds of natural disasters, so as to safeguard the public interests and people's life, property security, realizing society's normal operation and sustainable development.[2] The concept includes four aspects of specific meanings: Firstly, all types of natural disasters management. Secondly, all phases of natural disaster management.

---

[1] Yin Jie, Yin Zhane, Xu Shiyuan, Chen Zhenlou and Wang Jun. *Disaster Risk Theory and Risk Management Method Research*. Science of Disaster, 2009.

[2] Zhang Jiquan, Okada Norio and Tatano Hirokazu. *Comprehensive Natural Disaster Risk Management—Mode of Full Integrations and Strategic Choices in China*. Journal of Natural Diasater, 2006.

Thirdly, integrated natural disaster management. Fourthly, total risk management of natural disaster. ①

2.3 Invulnerability

In poverty research the World Bank's definition has strong representative, which considers invulnerability as the possibility of some risks met by an individual or a family, as well as the possibility of property loss or living quality drops under a level of social recognition because of encountering risk. The definition contains two aspects: one is the possibility and degree of encountering shock, and the other is the ability to resist shock. ② Dercon③ (2001) constructs an analytical framework of risk and invulnerability, bringing farmers' various resources, income, consumption and corresponding institutional arrangement (market mechanism, public policy, etc.) into a system. In the framework, farmers' risk origins mainly have three terms: Asset risk, income risk, welfare risk. ④

3. Literature Review

Generally speaking, the direction, theme and content of social science research have a high degree of correlation with social reality and its changes. Disaster risk management and poverty reduction theories are not exception. All these phenomena such as climate change, frequent disasters, huge loss, highlighted risk rising, associated poverty and so on gradually project the vision of academic research on to

---

① Okada N, Amendola. *A Research Challenges for Integrated Disaster Risk Management*. Presentation to the First Annual IIASA-DPRIM Meeting on Integrated Disaster Risk Management: Reducing Socio-Economic Vulnerability, at IIASA, Laxenburg, Austria (Aug 1-4, 2001), 2001;Okada N. *Urban Diagnosis and Integrated Disaster Risk Management*. Proceedings of the China-Japan EQTAP Symposium on Policy and Methodology for Urban Earthquake Disaster Management. November 2003, Xiamen, China. PP. 9-10.

② Han Zheng. *Vulnerability and Rural Poverty*. Agricultural Economic Problems, 2004.

③ Chambers. Robert, *Poverty and Livelihoods: Whose Reality Counts?* Environment and Urbanization, 1995(4).

④ Chen Chuanbo. *Farmers' Risk and Vulnerability: An Analytical Frame and Experience in Poor Areas*. Experience Problems in Agriculture, 2005.

theoretical propositions and practical work related to disaster risk management and poverty reduction.

3.1 Disaster, the relationship between disaster and poverty, between disaster risk management and poverty reduction

The coordination between disaster and poverty, disaster risk management and poverty reduction points to fundamental problems of research topics, and it certainly has fundamental value. To clarify the relevance, patterns, content and mechanism of the models the two relationships and present multi-dimensional features in the executional level of practice, operational level of policy, constructional level of theory, is of great priority for a premise and starting point of study. Zhang Xiao[①] described the general relationship between floods and droughts and rural poverty by giving examples of several typical areas. At the same time, he established an economic model of floods and droughts and its relation with the numbers of rural impoverished based on statistical data. Wang Guomin[②] elaborated on the positive correlation from four aspects between natural disaster and rural poverty. Natural disasters led to the rise of poverty incidence rates in rural areas. Sinking back into poverty is aggravated by natural disasters. Natural disasters have resulted in the underdeveloped construction of basic infrastructure. Natural disasters have restricted the economical development in rural areas. Ding Wenguang, Hu Lili, Wang Xiujuan[③] show that there is a high correlation between natural disasters and poverty by conducting quantitative analysis of natural disaster and net income of farmers in 87 counties in Gansu Province and 43 national poverty count. This correlation in the areas with ecological vulnerability, poor natural conditions, natural resources deficiency, economic and cultural backwardness poverty seemed to be

---

① Zhang Xiao. *Flood and Drought Disaster and Rural Poverty in China*. Rural Economy in China, 1999.

② Wang Guomin. *Natural Disasters in Agricultural and Rural Poverty Problems Research*. Economist, 2005.

③ Ding Wenguang & Hu Lili and Wang Xiujuan. *Natural Social Environment and Poverty Crisis Research in Gansu Province*. Environment and Sustainable Development, 2008.

more conspicuous. Xu wei① also analyzes the negative impact of disasters on poverty. However, he indicates that disaster prevention, mitigation and post-disaster reconstruction in practice working level which provides an advantageous opportunity for development-oriented poverty alleviation from practical working perspective. Furthermore, the regional distribution of disasters and poverty gradually attracted intellectual circle's attention. Li Xiaoyun, Tang Lixia, Chen Chongying② gave an example of Wenchuan earthquake-stricken region and discovered the phenomenon of coincidence between frequent natural disasters and high incidence of poverty. Huang Chengwei, Li Haijin③ also indicated the high degree of incidence in regional distribution between natural disasters and poverty.

The final results of three pairs of relationship (between disaster and poverty, between disaster risk and poverty, between disaster risk management and poverty reduction) are the framework of disaster risk management and poverty reduction, which is intensively embodied in the combination of strategies, mechanisms and tactics in the theoretical and policy level. Zhang Qi④, Wang Hao⑤ resolved strategtic analysis, institutional framework and specific strategies which combine poverty with disaster risk management by executing case analysis in disaster prevention and

---

① Xu Wei. *Wenchuan Earthquake's Influence on Poverty and Its Enlightenment for Disaster Emergency System*// Huang Chengwei, Lu Hanwen. *Poor Villages' Reconstruction after Earthquake in Wenchuan: Process and Challenges*. Social Science Documents Press, 2011.

② Li Xiaoyuan, Tang Liying, Chen Chongying. *Natural Disasters and Poverty in China*// Wu Zhong. *Rising Food Price, Natural Disasters and Poverty Reduction—The Second Social Development and Poverty Reduction Forum's Compilation of Data about China-ASEAN's Social Development*. China Financial and Economic Publishing House, 2009.

③ Huang Chenwei & Li Haijin. *Disaster Risk Management and Preliminary Research Frame of Poverty Reduction Theory and Method*. China Povery Alleviation, 2010.

④ Zhang Qi. *Mechanism Analysis about Combination of China's Disaster Prevention and Reduction and Poverty Alleviation Development—Take Poor Villages' Reconstruction after Earthquake in Wenchuan for Example*// Huang Chengwei, Lu Hanwen. *Poor Villages' Reconstruction after Earthquake in Wenchuan: Process and Challenges*. Social Science Documents Press, 2011.

⑤ Zhang Qi and Wang Hao. *Strategic Thinking about Combination of Rural Villages' Disaster Risk Precaution and Poverty Reduction in Special Type Regions—Take Highlands in Yushu, Jiangsu Province for Example*. The paper is included in the symposium.

mitigation through Wenchuan post-disaster reconstruction and special poverty-stricken areas which is represented by Yushu in Qinghai Plateau. And it tended to furnish a constructive direction and idea for relevant policies and practical work.

3.2 The influence of disasters on poverty

The influence of disasters on poverty is the fundamental and salient problem in disaster risk management, theories and practical research of poverty reduction. And it is also the most academic topic which is logically derived from disaster and its consequence. Li Xiaoyun, Zhao Xudong[1] had built sets of framework and methodology of peconstruction that focused on operation evaluation of social organization in post-disaster areas, loss assessment of family and its social support system, evaluation of farmers' livelihood system, assessment of social public service, evaluation of gender impact, assessment of food safety system effects and so on. All the above provided a referential guidebook in post-disaster evaluation for poverty-stricken areas and impoverished population. Based on documents and field research data, Li Xiaoyun, Tang Lixia, Chen Chongying[2] took Wenchuan earthquake as object of study and emphasized the damages accompanied with earthquake disasters in material capital of households, human capital, financial capital, natural capital and social capital. Given two dimensions from the historical and realistic perspective for the analysis of vulnerability, Huang Chengwei[3] conducted analysis framework for

---

[1] Li Xiaoyun & Zhao Xudong. *Social Assessment after Disaster: Framework Method*. Social Science Academic Press, 2008.

[2] Li Xiaoyun, Tang Liying and Chen Chongying. *Natural Disasters and Poverty in China* // Wu Zhong. *Rising Food Price, Natural Disasters and Poverty Reduction—The Second Social Development and Poverty Reduction Forum's Compilation of Data about China-ASEAN's Social Development*. China Financial and Economic Publishing House, 2009.

[3] Huang Chengwei, *Poor Villages' Restoration Planning Design and Implementation Prospect after Disaster in Wenchuan Earthquake of China* // Wu Zhong. *Rising Food Price, Natural Disasters and Poverty Reduction—The Second Social Development and Poverty Reduction Forun's Compilation of Data about China-ASEAN's Social Development*. China Financial and Economic Publishing House, 2009. Huang Chengwei. *Carrying out Theoretical Framework of Disaster Risk Management and Poverty Reduction's Theory and Practice Research*. China Poverty Alleriation, 2011.

infucence of the disasters on the poor and empirically investigated the influence of Wenchuan earthquake on poverty. In order to instruct and plan post-disaster relief work, State Council Office of Poverty Alleviation, which is responsible for the policy making, planning, organization and implementation of national poverty reduction, edited and published series of training books about disaster prevention and mitigation, post-disaster reconstruction and national development-oriented poverty alleviation. It roughly outlines the evaluation background, principles, procedures, index, methods and tools the impacts of disasters on poverty. And it also provides a framework for exploring the effects of disaster caused by poverty (edited by post-disaster reconstruction office for poverty-stricken villages — Council Office of Poverty Alleviation, 2010). Hu Jiaqi, Ming Liang[1] empirically studied the different effects of floods on poor households and non-poor households in human capital, physical capital, financial capital and social capital through a questionnaire survey as well as the changes in the revenue structure of different households under the influence of flood. Hu Jiaqi[2] conducted empirical investigation by using related theories of vulnerability for the poverty effects caused by droughts and discovered livelihood strategies of different households in response to droughts had great differences. Zhuang Tianhui, Zhang Haixia, Yang Jinxiu[3] analyzed the impacts of natural disaster on the rural poverty in minorities region from four levels and put forward four coping strategies for poverty alleviation by collecting and analysizing survey data of 67

---

[1] Hu Jiaqi, Ming Liang. *Based on the Research of Natural Disasters' Rural Poverty Effect—Take the Flood Survey in TL Village in the Southwest of Guangxi Province for Example*. Agriculture Science in Anhui, 2009.

[2] Hu Jiaqi. *Discussing Natural Disasters' Poverty Effect in Western Undeveloped Regions—Take the TP Village' Drought in Gansu Province for Example*. Agricultural Archaeology, 2010.

[3] Zhuang Tianhui, Zhang Haixia and Yang Jinxiu. *Research about Natural Disasters' Effects on Regions Inhabited by Ethnic Groups in Southwest—Based on the Analysis of 21 National Poverty-stricken nations*, 17 Villages. Rural Economy, 2010.

minorities villages.

Considering Luquan County in Yunnan Province as a place frequently striken by droughts, Zhang Guopei, Zhuang Tianhui, Zhang Haixia[1] carried out factor analysis of vulnerability of poor farmers under the influence of droughts by using related docucments and data. They believed that droughts are the main factors which gave rise to poverty and farmer's sinking back into poverty. Based on government statistical data, Xuwei[2] gave a comparative analysis of the differences in poverty conditions between pre- and post-disaster and depicted the effects of Wenchuan earthquake on poverty-stricken countries, poor villages and impoverished population. From the perspective of the responses of disasters-striken households, Li Xiaoyun, Tang Lixia, Chen Chongying thoroughly investigated the differences in impact degree, coping mechanisms and livelihood results for households with different economic status affected by different kinds of natural disasters. And it erected an associated schema between effects caused by poverty and economic status of affected farmers. In addition, it can promote the relevance and effectiveness of disaster response policy.

Sun Mengjie[3] estimated the influence of the scale and extent of rural poverty in Wenchuan earthquake by making use of FGT index measurement about poverty. He found that Wenchuan earthquake makes a substantial increase in the scale and degree of poverty. However, post-disaster subsidy policies helped nearly a quarter of households temporarily out of absolute poverty.

---

[1] Zhang Guopei, Zhuang Tianhui and Zhang Haixia. *Study about Natural Disasters' Effects on Poverty Vulnerability of Farmers—Take Drought in Luquan, Yunnan Province*, Journal of Jiangxi Agricultural University. Social Science Press, 2010.

[2] Xu Wei. *Wenchuan Earthquake's Influence on Poverty and its Enlightenment for Disaster Emergency System*// Huang Chengwei, Lu Hanwen. *Poor Villages' Reconstruction after Earthquake in Wenchuan: Process and Challenges.* Social Science Documents Press. 2011.

[3] Sun Mengjie. *Research about Natural Disasters' Effects on Rural Poverty in Disaster Areas—Take Wenchuan Earthquake for Example*, 2011 PHD. Thesis in China Agricultural University.

3.3 Poverty reduction strategies and measures response to disaster risk

Based on relationship model between natural disasters such as floods and droughts and rural poverty, Zhang Xiao[①](1999) discussed the external factors and coping measures in the poverty reduction and alleviation from multi-dimensional perspective of capital security, technological support and institutional indemnification. Wang Guomin[②] proposed a new idea for rural poverty alleviation named "four combinations" from the perspective of correlation between agriculture natural disasters and rural poverty. That is a combination of mitigation and poverty alleviation, a combination of national financial investment and agriculture insurance, a combination of poverty alleviation through education and poverty alleviation through immigration, a combination of anti-poverty and basic conditions for adequate food and clothing. Li Xiaoyun, Tang Lixia, Chen Chongying[③] grouped China's response mechanism to natural disasters into three levels of pre-disaster prevention mechanism, post-disaster emergency response mechanism and post-disaster reconstruction mechanism. And they also analysed the content of measures, current situation and responsible agencies or individuals in details. From the governmental level, Xu Wei[④] not only described mitigation effects between the cause of poverty and anti-poverty, but also depicted the necessity to pay the attention for impoverished population in the process of Wenchuan earthquake

---

① Zhang Xiao. *Flood and Drought Disaster and Rural Poverty in China*, *Rural Economy in China*, 1999.

② Wang Guomin. *Natural Disasters in Agricultural and Rural Poverty Problems Research*. Economist, 2005.

③ Li Xiaoyun, Tang Lixia and Chen Chongying. *Natural Disasters and Poverty in China* // Wu Zhong. *Rising Food Price, Natural Disasters and Poverty Reduction—The Second Social Development and Poverty Reduction Forum's Compilation of Data about China-ASEAN's Social Development*. China Financial and Economic Publishing House, 2009.

④ Xu Wei. *Wenchuan Earthquake's Influence on Poverty and its Enlightenment for Disaster Emergency System* // Huang Chengwei, Lu Hanwen. *Poor Villages' Reconstruction after Earthquake in Wenchuan: Process and Challenges*. Social Science Documents Press. 2011.

emergency relief and reconstruction.

Sun Mengjie[①] provided a diversity and targeted policy recommendations aiming at different social economic characteristics of various affected households, which publishing corresponding preferential policies to those areas insufficient capacity of self-development, introducing enterprises by adopting industrial poverty alleviation model for households in poverty-stricken areas with significant advantages of natural resources and labor resources, conducting targeted training such as construction skills, planting and breeding techniques and so on.

3.4 The contributions and deficiency of existing research

Through the review of existing research documents, we found that the main contributions and characteristics lie in the following aspects: (1)realistic enough it takes full account of social reality. Disaster risk management research gradually emerged as a main academic central issue in recent years. It originates from the reality of frequent disasters and serious consequences in China and internal needs for academic research. However, the overlapping of disaster-stricken areas and poverty distribution areas automatically achieves effective put-joint. It leads to explicit awareness of problems of disaster risk management and poverty reduction research in the origin of study. Accordingly it hits the crucial point of social reality and achieves organic link between risk management and academic research. (2)prominent empirical methods. Given intensive reality in the origin of research, the features of disaster risk management and poverty reduction study in research methodology are extremely distinctive. More empirical methods are applied. We have started detailed case analysis of earthquake disasters, droughts and floods especially the significant natural disasters that happened in recent years, and made a relatively positive response to current needs and policy requirements and achieved fairly good research effects. (3) Highlights

---

① Sun Mengjie. *Research about Natural Disasters' Effects on Rural Poverty in Disaster Areas—Take Wenchuan Earthquake for Example.* 2011 PHD. thesis in China Agricultural University.

of policy analysis. Corresponding with the above two features, practice-based policy analysis is the main type of research. And this kind of study is directed by problems and strategies. It uses the evaluation of the policy's timeliness, scientificness, appropriateness and beneficialness to the poor as fundamental content. And it also sets policy framework reconstruction and policy recommendation as basic direction. Therefore, it effectively plays a role of brain trust and makes due contribution to the supplement of disaster risk management and poverty reduction policy system.

The remaining shortcomings of existing research: (1) The preciseness of empirical methods need to be improved. The existing study on processing and analysis of empirical data (such as secon hand statistics material, questionnaire data, field interviews material, policy background) is difficult to achieve good results in the support and confirmation of statistics among different types and natures and it's hard to achieve organic link and adaptation in the direction of empirical analysis and theoretical view or the proposed policy recommendations. (2) The combination of policy analysis and theoretical research needs to be advanced. Disaster risk management and poverty reduction research has its own theoretical context. Its enchantment lies in explaining social phenomena or disclosing mechanism and internal logic behind problems. Policy-care theoretical study always thoroughly and effectively explores the subtle mechanism and consequence in policy system and its operational procedures. The organic combination in theoretical study and policy analysis mutually reinforces each other. But if the combination of surface alone or the correlation is not tight enough, there may be a significant damage to both sides. Existing research is not entirely the presentation of natural developmental stage, disaster itself involves a certain degree of contingency and emergency. Therefore, it faced serious conditions which lack of theoretical resources. It results in the instability of theoretical research of policy analysis. And it is difficult to obtain strong endorsement in the feasibility and applicability of policy instruments. What is more, policy intentions of theoretical investigation are excessively obvious; however, it will be confronted with the risk of being challenged in real explanatory power and logical preciseness. (3) The systemic and logic of research framework remains to be optimized. Disaster risk management and poverty reduction research is a systematic and highly logical academic issue which is related to multiple links such as motivation, manifestation, consequence, reaction and so on. And there is apparent connectivity in its all links. The existing research generally lack of systemic in the establishment of research framework. This kind of research strategy is so concerned about individual links that it is difficult to reveal the deep-seated operational mechanism and environment of each link.

4. The Construction of the Analysis Framework and its Logic

On the basis of critical reflection on existing research resources, the author has constructed a set of analytical framework of disaster risk management and poverty reduction theory as well as practice research, providing a solid basis for the study. This analytical framework focuses on two issues: Firstly, draw the outline of the logical relation between the research's concept and theme; secondly, draw the outline of the research's basic ideas and technical methods. Its goal is to keep proper balance and tension between three aspects: explanatory power, reality conviction and policy anticipation.

This framework mainly includes four parts:

The first part is the motivation. Explaining motivations and backgrounds of disaster risk management and poverty reduction theory as well as practice research from three concepts of disaster, disaster risk, poverty and two relations of disaster-poverty and disaster risk-poverty from aspects of theory and reality. What's more, the relational models are dimensional, not only disasters and disaster risk can lead to poverty or bring affected population into poverty at any time, but also poor population may adopt unsustainable developmental pattern due to lack of other alternative resources in the process of getting rid of poverty and becoming better off, which causes excessive use of natural resources and great destruction to natural environment, increasing frequency and possibility of disasters' occurrence.

The second is the performance. The two relationships of disaster-poverty and disaster risk-poverty exist in specific performance of three aspects. Firstly, disasters become one of the important factors that cause poverty, it is the so-called "poor because of disasters", "returning to poverty because of disasters", and the same degree of disasters have significant different influences on poor areas and developed areas. Secondly, overlapping of disasters and poverty in area distribution, disaster-prone areas and areas with a large poor population have a high degree of area overlapping. Thirdly, disasters have both natural attribute and social attribute with social attribute highlighted, thus have a close relation with regional conditions and human's subjectivity, which makes poor areas and weak population as well as the invulnerability exposed obviously in the disaster management.

The third is the consequence. Close relations among disasters, disasters risk and poverty result in poor population's invulnerability, what's more, the invulnerability is the high superimposition and accumulation of ecological fragility, economic fragility, social fragility. Ecological fragility refers that poor population often live in ecological fragile regions where natural environment is bad, natural disasters frequently occur, and it's difficult to keep proper balance between ecological protection and ecological destruction. Economic fragility refers to the low level, weakness and insufficiency of poor population in income and consumption economic development resources and conditions, market share and participation, etc. Social fragility refers to the institutional or mechanistic weakness of poor population in social capital, the right to speak, social participation, social exclusion, etc. In the background of above multiple fragilities' superimposition and accumulation, once poor population suffer from disasters, they all face quite disadvantegeous situations in different periods. Before a disaster occurs, the poor population who are characterized by low abilities of disaster prevention and reduction in many aspects such as production and life. When the disaster occurs, poor population's disaster-affected magnitude is heavy, resulting in the situation of "risk of being deprived of the only assets or surplus production". After the disaster, rehabilitation of poor population is difficult, it is difficult to recover to pre-disaster in a short time in the absence of strong external supports, thus the desire of getting rid of poverty and becoming better off becomes more far away.

The fourth is the response. It mainly includes three levels. Firstly, short-term measures, including the popularization of knowledge, training and drill of disaster prevention and reduction aiming at reducing vulnerability of poor population, rescues in the disaster of keeping an eye on poor areas and population, emergency management and social assistance, reconstruction after disaster as well as sustainable livelihood development and capacity building. Secondly, long-term policy, emphasizing sustainable development and comprehensive policy frame and long-term developmental strategies, along with key attentions for concentrated special poverty-stricken areas in the implementation of the new ten-year poverty alleviation and developmental programs. Thirdly, theoretical research, including distributed architecture analyses on areas where disasters frequently occur, areas where environment is fragile and areas that poverty alleviations focus on; the theory framework's construction of disaster risk management and poverty reduction, disaster risk management and international exchange in poverty reduction. In addition, we should pay attention to several combinations. the first is the level of main factors, joint actions and whole heartedly cooperation between the country and communities (and residents). As the country's representative, the government can be subdivided into central government and local governments (still can be further subdivided into basic governments), we should know differences of all levels of government's ideas, cognitions and actions, we should also pay attention to international knowledge's share and experience's borrowing. The second is the level of actual operation, inner logical relations and specific application in the research of the technological path

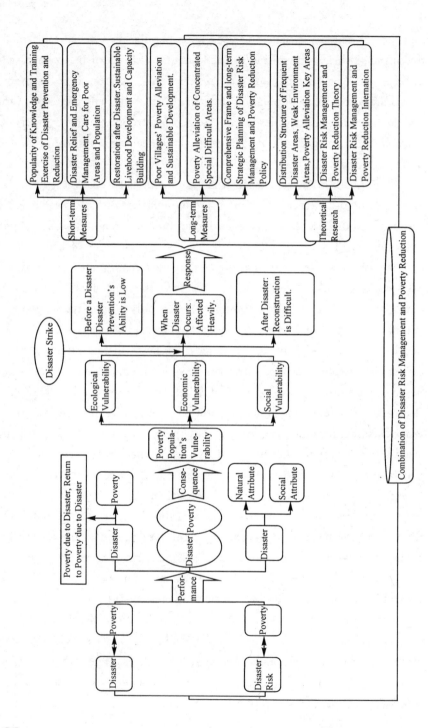

of "theory-practice-strategy-policy". This path is not single-directional, but double-directional and even networked. The third is the level of coping measures, according to acceptance or rejection of factors of the main demands and practical conditions, etc., attention should be paid to tension, connection and integration among short-term measures, medium-term strategies and long-term strategies/policies.

5. The Preliminary Application of Analysis Framework and Prospect of the Research

In order to give more sufficient, in-depth, comprehensive explanation for above analytical framework, the author took great natural disasters' coping measures in poor villages' reconstruction after Wenchuan earthquake for example, made empirical analysis on the key elements and core topics for disaster risk management and theories and practice of poverty reduction.

Firstly, in recent years, great natural disasters such as snow disaster, earthquake, droughts, floods, etc. all broke out intensively, From the perspective of poverty, in fact it indicates the changes of the reasons, characteristics and distribution of poverty. After the reform and opening up, because of sustaining, rapid economic growth and government's planned and organized mass development-oriented aid work, China's anti-poverty career has made great achievements, the problem of poverty has been alleviated largely. Poor population in rural areas has reduced from 250 million in the early period of reform to 14.79 million in 2007, poverty rate has dropped from 30.7% to 1.6%. However, with the reform advancing in depth and the deepening of economic and social development, the characteristics, causes, distributions of poverty are gradually becoming more complicated and changeable, anti-poverty work thus faces new problems and challenges. First of all, as for the causes of poverty, individual and family factors as well as cultural and psychological factors gradually occupy the dominant position. During the reform and opening up, China's farmers were in universal poverty, the poverty was mainly caused by institutional factors. Thereafter, reform of institution and policies promoted the rapid development of economy, poor population greatly reduced. For poverty problems, personal and family factors have gradually been highlighted. At the same time, social exclusion, the conservativeness of culture and psychology, ect. also become important causes resulting in poverty. Then, in the characteristics of poverty, individual poverty has considerably replaced overall poverty and become the main type of present farmers' poverty. Absolute poverty has been weakened, while relative poverty highlighted. With the organized, planned, wholesale poverty alleviation and development strategy fully implemented, overall poverty in rural areas has been weakened to some extent while individual poverty accordingly becomes apparent. The area distribution and population distribution of rural poor population increasingly decentralized and marginalized, the phenomenon of coexistence of centralized and decentralized appeared, many developed regions also have many poor farmers. Finally, in the distribution of poverty, the concentrated special poor areas where ecology is weak, living environment is bad are the concentrated areas of poor population, and also the main battlefield to realize the task of eliminating absolute poverty, becoming key areas of crucial poverty alleviation and development project in the new decade. "Central No. 1 Document" in 2010, a government work report in 2011, "12th Five-Year Plan" program outline as well as "China's rural poverty alleviation and development program (2011-2020)" that is drafting, all point out strategic changes in key areas of national poverty alleviation and development in the new decade from the poverty alleviation strategy, highlighting crucial poverty alleviation' strategic meanings in concentrated special poor areas.

Secondly, the high corelation between disasters and poverty embodied in causal relationship, regional distribution, basic attribute, etc. First of all, disasters and poverty are interwined and interchangble, transform each other between cause and effect in the causal relationship.

A Study on the Theories and Practice of Disaster Risk Management and Poverty Reduction: A New Issue

On the one hand, in the new period, short-term poverty or individual poverty such as poverty due to disasters, poverty due to illness, poverty due to going to school, etc. have increasingly replaced long-term poverty and universal poverty, becoming the main type of rural poverty. On the other hand, due to lack of other alternative resources, poor population may also adopt unsustainable developmental pattern during the process of getting rid of poverty and becoming better off, which will lead to excessive use of natural resources and great destruction of natural environment, thus increasing disasters' frequency and possibility. Then, in regional distribution, disasters and poverty have a high regional overlapping. As for the location, moderate and severe poor areas with large poverty coverage, deep poverty and great difficulties to overcome poverty are often backcountry where bad natural environment, limited developmental conditions and insufficient development opportunities. Due to their close relations with natural environment, natural disasters occures here with a high frequency, which is quite common for regional coincidence. According to statistics, poor areas' probability to suffer from serious natural disasters is five times as much as other areas. [1] In addition, from regional distribution of poverty-returning, the current regional characteristics of rural poverty-returning are very obvious as well, areas with deep degree of poverty returning and high probability of poverty-returning are often midwestern concentrated poor areas where natural environment is bad, natural disasters are frequent. [2] Finally, in the basic attribute, disasters and poverty's social attribute receive the same attention and have a close relation with practice and policy. No matter from the cause or consequence, disaster's social attribute gradually highlights, human factors have an increasing influence on disaster's happening and coping.

---

[1] Zhang Xinxin. *Task of our Nation's Poverty Alleviation and Development is Arduous in the Next Ten Years, Pressure of Returning to Poverty is still Great*. Xinhua net, 2010. http://www.gov.cn/jrzg/2010-12/21/content_1770521.htm.

[2] Jiao Guodong & Liao Fuzhou. *Perspective on Phenomena of Returning to Poverty in Poverty Relief*. Tribune of Study, 2001.

The unbalance and irrationality of social policies, rules and resources, etc. form different impacts on different affected areas. Correspondingly, the cognition of poverty gradually changing from locational, cultural, even economic perspective to social perspective; poverty's existence and emergence aren't caused by simple natural factors, but are more related to a society's developmental strategy, policy frame and management system, the phenomenon of poverty and the issue of poverty are not a simple economic problem, but also a social, political problem; poverty alleviation work cannot only be satisfied with improving bad natural environment for the locals or solving problems of food and clothing, but should also pay attention to person's development and social fairness and justice. ① Poverty belong to a "social disease", not a "culture disease".

Thirdly, what is closely related to the consequence of disaster, disaster risk and poverty is the vulnerability of poor population, and this vulnerability is the high superimposition and accumulation of ecological fragility, economic fragility and social fragility. In the background of above superimposition and accumulation of multiple fragilities, when poor population suffer from disasters, they always meet with fairly bad situations in different periods. Before a disaster occurs, poor population show the craracteristic of low ability of poverty prevention and reduction in production and life. During a disaster, poor population have a heavy suffering, appearing a situation of "having a risk of being deprived of the only assets or surplus production". After the disaster, rehabilitation of poverty population is difficult, it's difficult to recover to pre-disaster in a short time in the absence of strong external support, thus the desire of getting rid of poverty and becoming better off is more far away.

Fourthly, the stance of disaster risk management and poverty reduction is to explore coping measures from theories, practices and policies and so on. It mainly includes three levels: Firstly, short-term

---

① Xu Yong & Xiang Jiquan. *Antipoverty: Live, Service and Rights Protection.* Huazhong Normal University Proceedings(Humanities and Social Science Edition), 2009.

measures, including the popularization of knowledge, training and drill of disaster prevention and reduction aiming at reducing vulnerability of poor population, rescues in the disaster of keeping an eye on poor areas and population, emergency management and social assistance, reconstruction after disaster as well as sustainable livelihood development and capacity building. Secondly, long-term policy, emphasizing sustainable development and comprehensive policy frame and long-term developmental strategy, along with key attention for concentrated special poverty-stricken areas in the implementation of the new ten-year poverty alleviation and developmental program. This policy framework should realize efficient cohesion in real conviction and theoretical interpretation with both theory's foresight and practice's feasibility. In the content system, we should focus on factors of market mechanism, technological progress, economic growth, public financial expenditure, social fairness and justice, ect., Thirdly, theoretical research, including distributed architecture analyses of areas where disasters frequently occur, areas where environment is fragile and areas that poverty alleviations focus on; the theory framework's construction of disaster risk management and poverty reduction, disaster risk management and international exchange of poverty reduction. In addition, we should pay attention to several combinations. The first is the level of main factors, joint actions and whole heartedly cooperation of the country and communities (and residents). As the country's representative, the government can be subdivided into central government and local governments (still can be further subdivided into basic governments), we should know the differences of all levels of government's ideas, cognition and actions, and we should also pay attention to international knowledge's share and experience's borrowing. The second is the level of actual operation, inner logical relation and specific application in the research of the technological path of "theory-practice-strategy-policy", this path is not one-directional, but double-dimentional and even networked. The third is the level of coping measures, according to acceptance or rejection of factors of the main demands and practical conditions, etc. , attention should be paid to tension, connection and integration among short-term measures, medium-term strategies and long-term strategies/policy.

As for research's prospect, disaster risk management and poverty reduction's theory and practice have several aspects of expanding and orientation. Firstly, the synthetically use of multi-disciplinary theories, methods and skills, which complement and promote each other. This research covers resources and knowledge of environmental science, sociology, political science, economics, anthropology, human ecology and many other subjects, none of these subjects alone can do the research by itself. As a result, we must keep the openness and blending in subject category, and should set up a terrace for the intersection of disaster and poverty study, in order to form multi-disciplinary theory and expert resources. Secondly, it can help the organic combination and effective connection of ideas, knowledge, experience and skills which cross many field. Because this research is very comprehensive, systematic and inclusive, it contains three aspects of macroscopic, the medium, microcosmic research levels. Therefore, we should combine virtual and reality, unify of knowledge and action during the research, realizing the organic combination and effective connection of ideas, knowledge, experience and skills. Thirdly, multi-layered content and range of research includes at least the following aspects: from the environmental perspective, it includes environmental fragility and sustainability, biological variety and disaster-risk-management and ecological environment-risk-management of poor areas and so on; from the economic perspective, it includes economic structure, industrial structure, technological progress and disaster-risk-management and poverty reduction, ways of farmers' obtaining employment, income structure and disaster-risk-management and poverty reduction, financing, use and evaluation of risk-reduction and so on; from the cultural perspective, it includes developmental awareness, risk awareness, cultural traditions and risk-disaster-management of communities and farmers and minorities and disaster-risk-management and so on; from the political perspective, it includes social organization, capital and participation of disaster-risk-management, the social management system and operational mechanism and disaster-risk-management and so on; from the political perspective, it includes system factors, policy factors, civil rights, grass-roots democracy, disaster-risk-management and so on; from the comprehensive perspective, it includes gender, psychology, the building of information platform, government-market-society co-operational system and disaster-risk-management and poverty reduction and so on.

References:

[1] Chen Lizhong. Multidimensional Poverty Estimate and Decomposition of China in Transition Period[J]. Economic Review. 2008(5).

[2] Deng Tuo. A History of the Relief Work in China[M]. Beijing: Beijing Press. 1998.

[3] Ding Wenguang, Hu Lili, Wang Xiujuan. Natural Social Environment and Poverty Crisis Research in Gansu Province [J]. Environment and Sustainable Development, 2008(3).

[4] Gao Yunhong & Zhang Jianhua. Evolution of the Concept of Poverty[J]. Reformation, 2006(6).

[5] Gong Qianwen, et al. Relationship between Agricultural Natural Disasters and Rural Poverty——Based on Empirical Analysis of the Panel Data in Anhui Province[J]. China's Population. Resources and Environment, 2007(4).

[6] Guo Xibao & Luo Zhi. Discussing the Evolution of the Concept of Poverty [J]. Jiangxi Social Science. 2005(11).

[7] National Statistics Bureau's Rural Social Economic Survey Office. National Anti-poverty and Development Counties' Rural Absolute-Poverty Population is 17.63 Million[J]. Research World, 2004(6).

[8] Work Office of Poor Villages' Reconstruction after Disasters in the State Council Poverty Relief Office. Assessment Guidelines of Disasters' Effects on Poverty [M]. Beijing: China Financial and Economic Publishing House. 2010.

[9] Zhen Han. Vulnerability and Rural Poverty, Agricultural Economic Problems[J]. 2004(10).

[10] Hu Angang & Li Chunbo. New Poverty in New Century: Intellectual Poverty[J]. China's Social Science. 2001(3).

[11] Hu Angang & Tong Xuguang. China's Poverty Reduction Theory and Practice——in Qinghai's Perspective[J]. Journal of Tsinghua University(Philosophy and Social Science Edition), 2010(4).

[12] Hu Jiaqi & Ming Liang. Based on the Research of Natural Disasters' Rural Poverty Effect ——Take the Flood Survey in TL Village in Southwestern Guangxi Province for Example[J]. Agriculture Science in Anhui, 2009(28).

[13] Hu Jiaqi. Discussing Natural Disasters' Poverty Effect in Western Undeveloped Regions —— Take the TP Village' Drought in Gansu Province for Example[J]. Agriculture Archaeology, 2010(3).

[14] Huang Chenwei & Li Haijin. Disaster Risk Management and Preliminary Research Frame of Poverty Reduction Theory and Method[J]. China Poverty Alleviation, 2010(21).

[15] Huang Chengwei, Wang Xiaolin & Xu Liping. Poverty Vulnerability: Conceptual Framework and Measuring Methods [J]. Agriculture Technology Economy, 2010(8).

[16] Huang Chengwei. Theoretical Analysis of Combinations of Disaster Prevention and Reduction, Reconstruction after Disasters and Poverty Alleviation Development[M]//Huang Chengwei & Lu Hanwen. Poor Villages' Reconstruction after Earthquake in Wenchuan: Process and Challenges. Beijing: Social Science Documents Press, 2011.

[17] Huang Chengwei. Carrying Out Theoretical Framework of Disaster Risk Management and Poverty Reduction's Theory and Practice Research[J]. China Poverty Alleviation, 2010(23).

[18] Huang Chengwei. Poor Villages' Restoration Planning Design and Implementation Prospect after Disaster in Wenchuan Earthquake of China[M]// Wu Zhong. Rising Food Price, Natural Disasters and Poverty Reduction——The Second Social Development and Poverty Reduction Forum's Compilation of Data about China-ASEAN's Social Development. Beijing: China Financial and Economic Publishing House, 2009.

[19] Jiao Guodong & Liao Fuzhou. Perspective on Phenomena of Returning to Poverty in Poverty Relief[J]. Tribune of Study, 2001(1).

[20] Li Xiaoyun, Tang Lixia & Chen Chongying. Natural Disasters and Poverty in China [M]// Wu Zhong. Rising Food Price, Natural Disasters and Poverty Reduction——The Second Social Development and Poverty Reduction Forum's Compilation of Data about China-ASEAN's Social Development. Beijing: China Financial and Economic Publishing House, 2009

[21] Li Xiaoyun & Zhao Xudong. Social Assessment after Disaster: Framework Method[M]. Beijing: Social Science Academic Press, 2008.

[22] Li Xiaoyun. Previewing the Construction of Social Safety Net of Poverty Alleviation [J]. Forum Rakyat, 2010 (1).

[23] Noun Authorized Committee of the National Science and Technology. Geography Nouns[M]. 2ed. Beijing: Science Press. 2007.

[24] Shang Weiping & Yao Zhimou. Multidimensional Poverty Measurement Method Research[J]. Finance and Economy Research, 2005(12).

[25] Sun Mengjie. Research about Natural Disasters' Effects on Rural Poverty in Disaster Areas——Take Wenchuan Earthquake for Example[D]. Beijing: Economic Management Institute, China Agricultural University. 2011.

[26] Wang Yang. Victim Assistance and Rural Poverty[J]. China's Mitigation, 2007(4).

[27] Wang Guomin. Natural Disasters in Agricultural and Rural Poverty Problems Research[J]. Economist, 2005(3).

[28] Wang Xiaolin & Sabina Alkire. Multidimensional Poverty Survey: Meanings of Estimate and Policy[J]. Rural Economy in China, 2009(12).

[29] Wang Yalin. Study on China's Rural Poverty and Anti-poverty Measures [J]. Journal of State Administrative Institute, 2009(1).

[30] Xu Wei. Wenchuan Earthquake's Influence on Poverty and Its Enlightenment for Disaster Emergency System [M]//Huang Chengwei & Lu Hanwen. Poor Villages' Reconstruction after Earthquake in Wenchuan: Process and Challenges. Beijing: Social Science Documents Press, 2011.

[31] Xu Yong & Xiang Jiquan. Anti-poverty: Live, Service and Rights Protection[J]. Huazhong Normal University Proceedings (Humanities and Social Science Edition), 2009(4).

[32] Yin Jie, Yin Zhane, Xu Shiyuan, etc. Disaster Risk Theory and Risk Management Method Research[J]. Science of Disaster, 2009(2).

[33] Zhang Guopei, Zhuang Tianhui & Zhang Haixia. Study About Natural Disasters' Effects on Poverty Vulnerability of Farmers——Take Drought in Luquan, Yunnan Province[J]. proceedings of Jiangxi Agricultural University (Humanities and Social Science Edition), 2010(3).

[34] Zhang Jiquan, Okada Norio & Tatano Hirokazu. Comprehensive Natural Disaster Risk Management——Mode of Full Integrations and Strategic Choices in China[J]. Journal of Natural Disaster, 2006(1).

[35] Zhang Qi. Mechanism Analysis about Combination of China's Disaster

Prevention and Reduction and Poverty Alleviation Development——Take Poor Villages' Reconstruction after Earthquake in Wenchuan for Example[M]//Huang Chengwei & Lu Hanwen. Poor Villages' Reconstruction after Earthquake in Wenchuan: Process and Challenges. Beijing: Social Science Documents Press, 2011.

[36] Zhang Xiao. Flood and Drought Disaster and Rural Poverty in China[J]. Rural Economy in China, 1999(11).

[37] Zhang Xinxin. Task of our Nation's Poverty Alleviation and Development is Arduous in the Next Ten Years, Pressure of Returning to Poverty is Still Great. Xinhua net. 2010. http://www.gov.cn/jrzg/2010-12/21/content_1770521.htm.

[38] Zhuang Tianhui, Zhang Haixia & Yang Jinxiu. Research about Natural Disasters' Effects on Regions Inhabited by Ethnic Groups in Southwest——Based on the Analysis of 21 National Poverty-stricken nations' 17 Villages[J]. Rural Economy, 2010(7).

[39] Anthony Giddens. The Uncontrolled World[M]. Nanchang: Jiangxi People's Publishing House, 2002.

[40] Robert Chambers. Poverty and Livelihoods: Whose Reality Counts? [M]. Institute of Development Studies at the University of Sussex, 1995.

[41] Okada N. & Amendola. A Research Challenges for Integrated Disaster Risk Management[R]. Presentation to the First Annual IIASA-DPRIM Meeting on Integrated Disaster Risk Management: Reducing Socio-Economic Vulnerability, at IIASA, Laxenburg, Austria, August1-4, 2001.

[42] Okada N. Urban Diagnosis and Integrated Disaster Risk Management[R]. Proceedings of the China-Japan EQTAP Symposium on Police and Methodology for Urban Earthquake Disaster Management, Xiamen, China, November 9-10, 2003.

[43] Oppenheim C. Poverty: The Facts[M]. London: Child Poverty Action Group, 1993.

(Translated by Li Haijin)

Strategic Thinking on the Combination of the Disaster Risk Prevention with Poverty Reduction in Poor Villages in Special Areas

—A Case Study of Plateaus Areas Such As Yushu, Qinghai Province

Zhang Qi Wang Hao[*]

Abstract: The poor areas of special type should be attached great importance to in terms of disaster risk prevention. The strategic approaches applied to prompt the integration of disaster risk prevention and poverty reduction in the special areas are the following: reduce the scope and degree of disasters through disaster prevention and reduction; expand the effect and content of disaster relief through poverty alleviation and development; improve the combination of poverty alleviation and development and disaster prevention and reduction through the overall risk management of pre-disaster defense, inner-disaster emergency response and post-disaster reconstruction; strengthen the combination of disaster risk prevention and poverty reduction measures from the source through unified planning; improve relavant of the disaster prevention and reduction departments' integration efforts through the innovation of organizational mechanism;

---

[*] Zhang Qi is a professor at School of Economic and Resource Management, Beijing Normal University. Wang Hao is a doctoral student at School of Economic and Resource Management, Beijing Normal University.

promote the environmental protection and poverty reduction and development through the ecological financial measures in such areas; strengthen the constructional of productional facilities and technological service facilities through developing characteristic industries in such areas; elaborate the advantages of culture and customs in areas of special type through developing ethnic tourism.

Key words: Special Areas; Disaster Risk Prevention; Poverty Reduction

Since entering the new century, China's poverty alleviation has attained huge achievements. At the same time, it is facing new problems and challenges. Especially in recent years, under the influences of an increasing number of natural disasters in China, phenomena of poverty and poverty caused by disaster are particularly prominent in many areas with bad natural conditions. Therefore, combining disaster risk prevention with poverty reduction becomes a new mission of poverty reduction work for China in the new century.

1. The Characteristics in Special Areas and Poverty Caused by Disaster

According to the joint survey carried out by members of the State Council Leading Group Office of Poverty Alleviation and Development, it showed that poverty was more prominent in special areas in our country. For example, in 148,000 villages which implemented comprehensive poverty alleviation during the Eleventh Five-Year Plan period, population in absolute poverty and low-income population makes up 33% of the rural population. While the figure of key poverty alleviation counties in rocky mountain areas, desert areas, highland and cold areas, loess plateau, high prevalence of endemic disease areas, minorities with less population areas and "direct areas" (direct transition from primitive society to socialist society) and the 42 border counties was more than 40%.

1.1 The disaster risks of plateau areas

Plateau areas in China mainly include the loess plateau, Qinghai-

Tibet plateau, Southwest karts' plateau areas and so on. Among them, the loess plateau and Qinghai-Tibet plateau are the key supported regions of the national poverty alleviation. Due to the unconsolidated parent materials, loess plateau suffers from the water erosion and cutting and these areas formed a morphology of broken land surface. Instability, windiness and drought are the main characteristics of the climate system in these areas, annual precipitation is around 400 milliliters, and spatial and temporal distribution is extremely uneven. This landscape and climate cause extremely serious soil erosion, poor cultivated land quality and low agricultural production capacity in the loess plateau. Qinghai-Tibet plateau's terrain is high and volatile, the cold climate is the main restrictive factors of agricultural production, and it appears as the tundra landscape meadow steppe. The annual precipitation is less than 400 milliliters, and it changes greatly in different years. Sometimes precipitation in drought years is less than half of that in the usual years. Shortage of heat makes the growing season shorter, crop and forage species extremely limited and the resistant capability against natural disasters such as drought, low temperature and disease is weak, and output is low, soil erosion and soil desertification is severe. It should be said that many factors restrict the development of poor areas at high altitude, but the most important factors are as following: The first is the historical reasons. Due to the harsh environment, it is sparsely populated since ancient times, development at high altitude is lagging behind. Since 1950, large scale virgin land, exploration and the indiscriminate tree felling and other anthropogenic factors resulted in particularly serious soil erosion, difficult recovery of vegetation and serious lack of water resources in the loess plateau at high altitude. Second is the geographical factors. As is mentioned before, poor geographic conditions, the insufficient construction of irrigation and water conservancy facilities resulted in farmers' weak capability against natural disasters. Third is the transportation factors. Underdeveloped transportation leads to the high of production costs and low production efficiency. Forth is the cost. Because of the harsh geographical conditions at high altitude, infrastructure reconstruction is difficult and expensive.

1.2 Particularities of the Yushu

Yushu in Qinghai province is a typical plateau region of China, and it is also one of the ethnic areas where many impoverished population are distributed.

Table 1 Particularities mainly reflected in Yushu, Qinghai province

	Content	Particularity
1	Geographical climate	Elevation of 4,000 meters above the sea-level, frigid and anoxic, great temperature disparity between day and night, short frost-free season.
2	Eco-environment	Vulnerable alpine meadow ecosystem, sparse vegetation, easy to loss soil and water, suffering damage from rats frequently, weak resistance.
3	Transport facilities	The density of highway network is low, and road conditions poor; there are only 2 main traffic roads with long traffic distance and high cost.
4	Infrastructure environment	Narrow terrain, small construction platforms, big difficulties in organization and coordination, large mechanical machines are difficult to go out and in.
5	Construction materials	Lack of building resources, main building materials relying on shipping from outside, high cost, lack of technical personnels and construction workers
6	Economic development	Economic underdevelopment, single industrial structure, weak financial resources, low income levels of peasants and herdsmen.
7	National conditions	Dominated by Tibetans, and the minority population is big.
8	Social culture	Under the far-reaching influence of Tibetan Buddhism, many monks.

Source: The material of the table were accumulated from all kinds of information.

Strategic Thinking on the Combination of the Disaster Risk Prevention with Poverty Reduction in Poor Villages in Special Areas

As is shown in table 1, compared with the general conditions of other plateaus, Yushu in Qinghai province has more particularities mainly embodied in following aspects: first, Yushu is located in the Northern Qinghai-Tibet plateau with harsh natural conditions. With an average altitude of above 4,000 meters, its weather is frigid and anoxic, and its temperature disparity is big between day and night, and the frost-free period is very short. The effective construction period is only 5 months a year which brings great difficulties to the development of this region. Second, eco-environment is vulnerable in Yushu. Most regions are of extremely vulnerable Alpine meadow ecosystem with short vegetation growing season, easy loss of soil and water and weak capability against external influence, and therefore it is extremely difficult to recover after damage. Third, transport facilities in Yushu fall behind compared with other regions. Huge disaster area, low density of highway network, poor road conditions; there are only 2 main traffic roads (National Highway 214 and provincial Highway 308) leading to long traffic distance and high cost. Fourth, environmental infrastructure construction in Yushu in Qinghai province is poor. Narrow terrain makes the construction platforms small, big difficulties in organization and coordination, logistics support capability is weak. Fifth, Yushu lacks building resources. Main local building materials rely on external inputs. Professionals in design, construction and management are in great shortage, and professional construction teams adapting to plateau job are also in shortage. Sixth, economic foundation in Yushu is weak. The disaster area is characterized by extensive grassland stockbreeding, single industrial structure, limited local governments' financial resources, low income of farmers and herdsmen, wide-ranged poverty, leading to weak self-recovery capability. Seventh, Yushu is inhabited by minority nationalities. The percentage of minority in the total population is more than 97% in the areas, and the Tibetan proportion reaches 93%. Tibetan Autonomous Prefecture in Yushu has a rich national cultural heritage and distinct geographical characteristics. Eighth, the influence of religion is strong in Yushu. This area is a gathering place of the many sects of Tibetan Buddhism; there are many temples, monks and religious believers, thus the religious influence is far-reaching.

To sum up, as a gathering area of minorities in alpine region of Qinghai plateau area, Yushu has very typical particularities. considering these particularities, on the one hand, poverty caused by natural geographical conditions is difficult to eradicate. On the other hand, disaster relief and disaster prevention measures can not be carried out smoothly after the occurrence of natural disasters. Therefore, due to the frequent natural disasters in this region, poverty caused by disasters has gradually stood out in recent years.

1.3 The status of poverty caused by disaster in Yushu in Qinghai plateau

Because of the special geographical environment, Yushu is one of the special areas where natural disasters often occur. Since 2006, the rate of poverty-returning has increased gradually to 25% and come to 50% to 60% in heavy disaster year. Poverty-returning population caused by the snowstorm in 2008 reached 39,561 people (see table 2), and the number is 10 times more than or even dozens of times more than that of the adjacent year. In particular, the Yushu earthquake in 2010 expanded the poverty rapidly, the rate of poverty-returning caused by disaster increased significantly, incidence of poverty rose from pre-disaster 34% to above 71%. ① The nationwide additional number of people trapped in poverty is 60,000, and total 200,000, accounting for 65% of all peasants and herdsmen, nearly the whole Yushu County returned to poverty. In addition, poverty level has further deepened in Yushu, public service capacity has declined significantly, the production and livelihood of impoverished masses suffered severe damage, and the economic and social development in Yushu endured a serious step of

---

① Development-oriented Poverty Reduction Office in Qinghai Province. *The Review of Achievements of Development-oriented Poverty Reduction Program in the Eleventh Five-Year Plan in Qinghai province.* Journal of Qinghai Peasants, November in 2010.

backwardness.

Table 2　The changing status of impoverished population in Yushu (unit: person)

year	impoverished population	stable population out of poverty	poverty-returning population caused by disaster
2006	93,558	——	757
2007	77,706	10,297	3,781
2008	104,914	12,353	39,561
2009	96,389	9,677	1,152

　　Source: according to the Yushu's National Economic and Social Development in 2006, 2007, 2008 and 2009.

　　Moreover, a large part of Yushu is located in Sanjiangyuan Nature Reserve. Since the establishment of Sanjiangyuan Nature Reserve, the government has implemented a series of ecological protection projects and measures such as returning land for farming to forestry, ceasing grazing and grass planting, stopping gold mining, and limiting the collection of Chinese herbal medicine. The local fiscal revenue had significantly decreased, For example, because of banning gold mining caused an annual loss of 20 million yuan. After stopping husbandry in the grassland, herdsmen's income level declined, coupled with only 5.5 kg grain subsidy per day, which made pastoral production and life hard to sustain, the risk of poverty-returning increased. Besides, to ecological immigrants, some pastoralists' follow-up industries could not be implemented. With the rising cost of living, limited subsidy and food supply, these pastoralists' livelihoods were actually not solved, especially for the pastoralists who have no skill except grazing, they are very prone to returning poverty and the poverty reduction progress is hard to last. Therefore, combining the disaster risk prevention with poverty reduction has become an inevitable choice to prevent disaster and reduce poverty in areas of special types.

2. Analysis of the Necessity of the Combination of Disaster Prevention and Reduction with Poverty Alleviation

Disaster risk is one of the main causes of poverty at this stage in China. At present, impoverished population mostly live in areas where the natural condition is harsh and economic is backward. These areas have low production capacity, excessive population growth, weak production technology, and insufficient funding. The ability to improve the economic value of resources and promote the economic transformation of resources is not strong, and economic development is slow. It should be said that, poverty and backwardness make people deforest for farmland, plant on steep slope and so on to obtain sufficient resources to maintain life, and lack of awareness of ecological resource protection leads to ecological degradation, soil erosion, frequent natural disasters, which brings about more serious poverty and forms a vicious circle as "poverty-ecological destruction-natural disasters-further poverty".

2.1 Disaster prevention and reduction is a priority task of poverty reduction program in special areas

Since entering the new period, with the gradual global warming, extreme weather happened constantly. Increasingly serious droughts in northern China, floods, landslides and debris flow caused by torrential rain in southern and western regions have resulted in increasing severe crop failures and life security can not be guaranteed. Because the less precipitation and more evaporation, water deficiency is in 300 milliliters to 650 milliliters, droughts is highlighted in loess plateau area. At the same time, due to intensive rainfall in autumn, low vegetation cover age rate and loose soil gives rise to serious soil erosion and ecological environmental deterioration. In addition, ecological vulnerability and drought is often accompanied with poor transportation, laggard culture and underdeveloped market economy. Various intertwined factors deepens poverty levels and increases the difficulties of poverty reduction in special areas. All above are the leading reasons for the prolonged poverty and backwardness in these areas. Therefore, preventing disaster risk and reducing the number of poverty-stricken and poverty-returning population after the disaster have become priory work of poverty alleviation and development in the special areas. Breaking down the thinking patterns of simple poverty reduction and suiting measures to local conditions to establish the disaster prevention and reduction and post-disaster reconstruction policy mechanisms will play a significant role in promoting the poverty reduction.

2.2 Disaster prevention and reduction is an effective way to prevent frequent poverty in special areas

In a long time, disasters relief was regarded as the core work. However, it turns out that only relying on disaster relief was not a fundamental approach of reducing poverty. Production is often destroyed absolutely when the areas are stricken by natural disasters, and economic development often suffer huge loss and damage. Natural disasters not only constrained the economic development, but also destructed and offset the long-term development achievements. Especially for areas of special types, disasters frequently happened, which caused larger damage. Therefore, in order to eradicate poverty, these areas should develop production and improve living conditions as well as actively carry out disaster prevention and reduction work to reduce and eliminate the negative impact of disasters. Considering the long-term effects of disasters, disaster reduction has long-term effects in poverty reduction. Combining poverty alleviation and development, people's living standards and disaster loss easing would increase the benefits of poverty reduction and guarantee the stable and sustainable economic development, thus speed up the pace of poverty reduction. For example, in Qinghai province, from year 2001 to 2008, Central Committee injected 5,209 million yuan into Qinghai province as alleviation project funds. It gave full support to Qinghai province's poverty reduction cause through policy, financial and project support, proposed to break the boundaries of key counties and poverty-stricken villages in poverty alleviation and development, implement contiguous development and comprehensive management. Considering the special situation that Tibetan areas in Qinghai province are located in the alpine area, Central Committee appropriately relaxed the poverty standard. It increased the inputs of poverty reduction funds, work relief and credit funds; strengthened the efforts of immigration, improved the implementing efforts of "Dew Plan", and strengthened the practical skills training for poor farmers and herdsmen. These specific measures directly reduced the impoverished population in Qinghai province from 1,976 million in the year of 2000 to 0.679 million by the end of year 2008, 1.3 million people got rid of poverty during 8 years.

2.3 Disaster prevention and reduction provides important support to enhance production and living standards in areas of special type

In the economic development of poor areas, disaster prevention and reduction not only shows the functions of disaster prevention and risk reduction, but also shoulders the important task of protecting ecological environment and agricultural production environment. Due to the fragile ecological environment and frequent natural disasters in poor areas, taking economic development alone as the main means of poverty alleviation and development would suffer the negative effects of the harsh natural environment. Therefore, disaster prevention and reduction can effectively guarantee the production and living environments in areas of special type, and it can create objective conditions to improve the standard of living through program development and whole village promotion programs. In addition, because the happening of natural disasters is closely related to the human's destructive behaviors toward ecological environment, realizing the importance of ecological construction through the sense of sustainable development is necessary; containing the effect of poverty reduction by sacrificing ecological environment is infeasible. The combination of disaster prevention and reduction with poverty alleviation and development is the important means to achieve sustainable development in poor areas. Only by combining these two aspects together, can the poor areas achieve the dual aim of environmental optimization and production development at the same time, and then the true enhancement of the socio-economic development can be advanced in poor areas.

3. The Strategic Thought on the Combination of Disaster Risk Prevention with Poverty Reduction in Special Areas

The crucial point of the combination of disaster risk prevention with poverty reduction is the innovation in mechanisms, which means the series of mechanism innovation of process management from planning, management, organization, resource allocation to the manpower arrangements. It is said that the mechanism is a set of scientific management methods and systematic working law, and it is also a theoretical summary of practical experience. According to the academic views, due to the consistency of the theory of disaster risk management and the theory of poverty reduction in the temporal dimension, and the similarity in reducing vulnerability of impoverished population. Therefore, in cases of the high overlapping of the disaster-prone areas with the poor areas of special types, disaster prevention and reduction work and poverty alleviation and development can form a mutual integration and virtuous cycle system (see Figure 1). In this mutual integrated cycle system, if a disaster occurrs, the combination of disaster and poverty-returning factors will cause losses in poor areas(see the part under the dash); after the disaster occured, the combined mechanism of disaster prevention and reduction work and poverty alleviation and development plays an important role in restoring the basic facilities and life order in poor areas at first, then enhance self-development ability, thus improving the level of disaster prevention and disaster reduction (see the part above the dash).

Therefore, it's necessary to form an effective implementation mechanism to achieve the strategic goals of combining disaster risk prevention with poverty reduction together. That is to say, from a strategic perspective, constantly promoting the establishment of the disaster risk prevention and poverty reduction mechanism is the prerequisite for implementing the established goals in areas of special types.

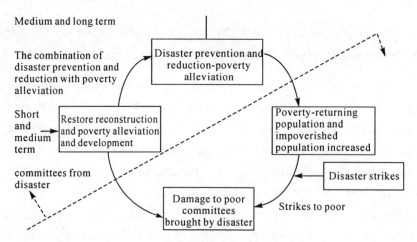

Figure 1 The interactive mechanism in the combination of disaster prevention and reduction with poverty alleviation

3.1 Reducing the scope and extent of poverty caused by disaster through poverty prevention and reduction.

In areas of special types, disaster prevention and reduction has been the important task in urban and rural development. Reducing the frequency of the occurrence of natural disasters effectively, and good early prevention and warning are effective measures to reduce disaster impact. For the current disaster prevention and reduction work, construction of disaster prevention and reduction program is the primary means against natural disasters. The main measures to reduce disasters are as follows constructing embankment dam in flood-prone areas, building huge mountain protection construction and planting trees in the key debris slope region, implementing relocation and constructing houses in geological disaster-prone areas. Combining disaster prevention and disaster reduction with poverty alleviation and development is to adopt means such as work relief, directional contract, authorized construction possibly in the working process, and letting people in poor areas take part in the construction program. In this process, the local government should take into account skill training toward people in poor areas, and let them master simple working skills. It not only constructs disaster prevention engineering, but also solves the problems related to people's livelihoods and the career skills development in poor areas. Training and exercises enhance the professional abilities of impoverished population, and lay a foundation for future development-oriented poverty reduction programs. For example, Nanjiang County of Sichuan Province has implemented relief work mechanism in the disaster prevention and poverty alleviation and development in a wide range of areas. The Relief Work Office is guided by Development and Reform Bureau of Nanjiang County, which masters the funding for relief work from superior governmental departments, and possesses the convenience because Development and Reform Bureau coordinates other functional departments. Through joint meetings, poverty reduction department guides relief work funds to be allocated into poor villages, bundles the funds with other items, and enable the interaction between different departments. Professional function departments participate in joint meetings, coordinating with the programs implementation, ensure the standards of programs. Implement joint-office system during the implementation of relief work; adhere to the "three common principles" of participating in program planning design, program management, and quality supervision. This system exerts resources mobilization ability of modified council, makes use of project aiming and project design ability of poverty alleviation and development, and takes advantages of other professional departments' technologies, thus ensure the excellent completion of various engineering and obtaining a good effect.

3.2 Expanding the content and limitation of disaster relief through poverty alleviation and development

In areas of special types, the disaster relief work becomes a heavy task once the disaster happens. The current disaster relief work mainly involves contents at two levels: one is the resettlement and relief of victims, which can ensure the basic livelihood needs of victims and is an important task of disaster relief in the short term; the other one is the living arrangements and future planning for victims, which is the key aspect of the long term relief after disaster. In this area, the poverty reduction work can be combined with the disaster prevention and reduction work, The issues of the future livelihood of victims should be considered from a sustainable developmental perspective, and should be combined with the poverty alleviation and developmental program to attain the long term protection of production and life. As previously mentioned, natural disaster is a major cause of poverty in areas of special types. Therefore, post-disaster relief to victims would certainly be combined with poverty alleviation and development; it could achieve the win-win effects of poverty relief and poverty alleviation and development through the unified planning of the arrangement and the future development. For example, after the Wenchuan earthquake, the counterpart donating aid from Beijing to Shifang not only helped the house recovery of Shifang people, but also concentrated on building the integrated 12,000 square meters' farmhouse. The farmhouse had functions such as catering, lodging and fishing. The ownership and all income were owned by the whole people of Yujiang village. The villagers were shareholders as well as employees. The people's livelihood attained a long-term protection. At the same time, Beijing's aid was also invested in the Jingshi industrial park. It acted as a bridge and attracted 14 companies to industrial park with a total investment of 2.263 billion yuan, and another 12 reached agreement of intented projects with a total investment of 895 million yuan. Therefore, after the occurrence of disaster in areas of special types besides short-term rescources, housing and other aids, it also needs to consider the characteristics in areas of special type, take industrial and strategic measures into account to guarantee the long-term poverty reduction to victims, combine short-term relief with long-term poverty reduction, and expand the content and limitation of disaster relief through poverty alleviation and development.

3.3 Promoting the combination of disaster prevention and reduction with poverty alleviation and development through the three-phase disaster risk management in the whole process

Disaster risk management theory proposes that the disaster risk management can be divided into three stages: pre-disaster prevention, disaster response and post-disaster rehabilitation and reconstruction. Furthermore, these three stages are three cycles that are mutually progressive and dependent on each other. Especially in post-disaster rehabilitation and reconstruction, the scientific and rational reconstruction plan can effectively improve the pre-disaster defense level in areas of special types, thus lay a good foundation to reduce disaster risk. Therefore, the combination of disaster risk prevention with poverty reduction is to combine risk disaster prevention work with poverty reduction efforts. The implementing of this combination can be divided into two different stages: pre-disaster and post-disaster. First, in the pre-disaster stage, poverty relief and alleviation is incorporated into the everyday disaster prevention and reduction work, and the combination of disaster prevention and reduction with poverty reduction measures can be achieved. Second, disaster prevention and reduction is incorporated into poverty alleviation and development, and effects of long-term poverty reduction are taken into account in the disaster reduction and relief operations after the disaster (see in Figure 2)

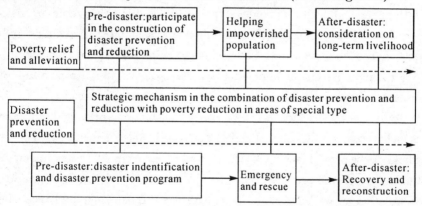

Figure 2 Effective mechanism in the combination of disaster prevention and reduction with poverty reduction

So, for areas of special type, strengthening construction of disaster prevention and reduction system and integrating disaster reduction capacity are effective measures to prevent the returning to poverty. For example, before a disaster, it is of great significance to enhance the construction of disaster reduction engineering and prevention of geographic phenomena such as earthquake, geology, and meteorology in the plateau areas; strengthen the professional monitoring system of floods in areas near oceans, lakes and rivers and improve the early-warning capacity construction of monitoring and forecasting debris flows and landslides in mountainous areas such as high mountains, valleys and majestic mountains. At the same time, it is important to strengthen basic mapping work, recover the construction of mapping benchmark infrastructure, gather basic geographic data and build geographic information public service platform. During the disaster, the government should take full advantage of funds and materials for disaster prevention and reduction and poverty alleviation and development, provide the timely aid and real-time help to victims. After the disaster, it is necessary to priority to the problems of sustainable poverty relief, maintain production and construction on victims' livelihood, promote the combination of disaster prevention and reduction with poverty alleviation and development throughout the whole process.

3.4 Strengthening the combination of disaster risk prevention with poverty reduction measures from the sources through unified planning

Planning mechanism plays a leading role in the combination of disaster prevention with poverty reduction in areas of special type. It is not only the conjunction point between disaster prevention and reduction as well as poverty alleviation and development, but also is the platform of the whole mechanism. Planning and management can comprehensively arrange the external integration management system and internal vitality stimulation as well as capability cultivation mechanisms, and reconcile the relationships among different works such as regional development, recovery reconstruction and poverty alleviation and development, etc. More importantly, it can combine long-term disaster prevention and reduction work with poverty alleviation and development in areas of special type, consider the effect of climate changes and disaster management on poverty alleviation and development in such areas, comprehensively arrange the design, construction and operations maintenance of various disaster prevention and reduction programs, deploy and arrange the local residents from the views of disaster prevention and reduction and poverty alleviation and development in the maximum degree. It's necessary to set up a high-level planning organization according to the existing working experience, which is under leadership of local planning department to integrate the two separate plans and form the overall plan of disaster prevention and reduction and poverty alleviation and development and guide the work of the whole areas. For example, the State Council has compiled a general plan for the recovery reconstruction of Yushu Earthquake in June, 2010. The plan adheres to put people first, respect nature through comprehensive arrangement and joint promotion, based on the reality of economic, society, natural geographic, ecological environment, and religious culture in Yushu, and learn from the successful experience of disaster reconstruction in Wenchuan earthquake. Therefore, the plan effectively combines reconstruction with strengthening the protection of Sanjiangyuan Nature Reserve, promotes economic and social development in minorities areas, improves people's production and living conditions, implementes poverty alleviation and development, and keeps the minorities characteristics and geographical features. In the end a socialist new Yushu with better ecological environment, distinctive characteristics, economic development, security and harmony is constructed. For example, the new plan toward regions and urban areas took full account of the factors such as natural geographical conditions and geological disaster-prone in Yushu, combined the national protection plan in Sanjiangyuan Nature Reserve, divided the planning areas into ecological reserve, modest reconstructive areas and comprehensive developmental areas. Among them, ecological reserve mainly included the core zones, buffer zones, experimental zones and zones that are suitable for developing husbandry in Sanjiangyuan Nature Reserve and Longbao Nature Reserve. Modest reconstructive area mainly included modest agriculture and husbandry zones and relatively intensively populated zones in the northwest and east of Yushu. In this area, it should implement reconstruction in-place of villages and towns, stabilize population, control the degree of agriculture and husbandry and town construction, recover the ecosystem functions gradually, focus on implementation projects such as returning grazing land to grassland, management of degraded grassland, conservation on wetland and wildlife, develope eco-tourism and other special industries appropriately. Comprehensive developmental area mainly refers to the 27 towns where are the locations of town governments, and population should be appropriately gathered.

3.5 Increasing the integration of departments related to disaster prevention and poverty reduction program through innovative organizational mechanisms

Organizational mechanisms form a joint force through the integration of disaster prevention and reduction department, poverty reduction department and other functional departments, which can enhance governmental departments' coordinative capacity, strengthen the cooperation between mobilization of resources and the use of resources, and ultimately achieve the organic combination of disaster prevention and reduction with poverty reduction and development on collaborative levels. It should be mentioned that organization mechanism acts as a link in the combination of post-disaster reconstruction and poverty alleviation and development. On the one hand, organization mechanism is the organizational guarantee of the implementation of planning and management; on the other hand, it is the precondition of resource integration configuration. According to the current situations of disaster prevention and reduction and poverty alleviation and development in areas of special type, these two aspects can be incorporated into a single department named Disaster Prevention and Poverty Reduction Office. The chief local leadership should lead and command this office and mobilize other related departments to implement relevant work as much as possible to ensure the combination of high degree of flexibility with basic principles in related departments.

The mechanism of coordinating resources can integrate different resources with the same characteristics together to achieve normalization and integration of disaster prevention and reduction, poverty alleviation and development and other resources in management and utilization, regulation, etc. Resources integration is a specific way to implement the external integration management mechanism, it is guided by the overall planning, and good departmental organization and coordination is the preconditions of resources integration. In the combined mechanism of disaster prevention, disaster reduction and poverty reduction, the resources coordination includes three main aspects. The first aspect is the overall management of funding, including central and local special funds for disaster prevention and reduction, the fiscal poverty relief funds, loans from financial institutions, relief work funds, etc. The second aspect is the overall management of material, including various types such as disaster prevention materials, stockpiles, poverty relief materials, and production and living materials. The third aspect is the overall management of human resources including various types of special talents, technological personnels and labors.

3.6 Promoting the protection of ecological environment and development of poverty reduction through eco-finance in areas of special type

Eco-finance is raised because of issues such as numerous nature reserves in areas of special type and areas where the protection of ecological environment restricts economic development. Financial compensation is an important measure to provide capital support and guarantee for the development of those areas. Many countries and regions have tried such measures currently, carbon sinks trading and natural disasters securities. For example, carbon sinks trading is an exchange process that some regions reduce the emission of carbon dioxide or raise the absorption of carbon dioxide, and then resell these targets of surplus carbon emissions to other regions to offset the emission reduction task in such areas. Since most poor areas in China are located in regions with better ecological conditions and weak industries, particularly in remote mountainous areas and ecological reserves. In order to maintain ecological diversity, these areas are unable to develop industries in due to national policies and regional plannings, then pose massive restriction on the path of getting of poverty. Therefore, the basis and criteria of carbon sinks trading are fully equipped. Moreover, from the regional point of view, carbon sinks trading has large prospects of application and implementation; the reason is that it is good for balancing regional development, and is the primary means to support underdeveloped regions through effective mechanisms. For example, in Yushu area, there are many national and provincial key natural protectional regions such as Sanjiangyuan Nature Reserve and Longbao Nature Reserve, and abundant natural resources such as rich grassland and water. In particular, as the state has gradually strengthened protection on construction and support of natural ecological protectional reserves, and increasingly devoted efforts to project construction such as forest conservation, comprehensive rehabilitation of small basins, which will speed up the reconstructions such as water conservation areas and nature reserve management facilities, and repairing ecological system function have gradually become popular. In particular, the recovery of grasslands and pastures is mostly achieved by ways such as strengthening grassland enclosure, executing regional rotation grazing, delaying grazing and banning grazing, and constructing artificial grassland in areas where conditions are available. These measures further limit the pastoralists' main way to promote economic development by grazing. Therefore, there are better conditions for implementing new eco-financial model in areas of special type.

3.7 Strengthening the construction of production and technical service facilities through developing characteristic industries in special areas

Development of special industry is closely related to the construction of production facilities and technical service facilities in areas of special type. Due to the limited natural geographical conditions, many kinds of industries which can be developed under normal circumstances are not applicable to areas of special type. Therefore, developing characteristic industries suitable to local conditions in special areas is an effective approach to implement disaster reduction and poverty alleviation and development. For example, due to the high altitude and rich alpine pastures, Yushu is suitable to develop the ecological husbandry. It means developing professional cooperative companies of peasants and herdsmen, intensifying rejuvenation of yak, selected breeding of Tibetan sheep and promotion of dzo, promoting the construction of breeding system, breeding base and breeding areas, and rehabilitating the reconstruction of livestock shed and grass stored shed. At the same time, considering the great of intensity sunlight at high altitude, it is more suitable to promote the new varieties of crops and new technologies such as barley, potato, Tibetan medicine, etc. Therefore, seed-breeding bases of barleys and potatoes can be constructed; efforts should be made to vigorously develop the agriculture with local characteristics, develop the production of potatoes, highland vegetables and Tibetan medicine and construct high standard energy-saving solar greenhouse in suitable areas, gradually raise the level of vegetable self-sufficiency. On this basis, it can develop farm and livestock products of the plateau characteristics, actively promote ecological agriculture and husbandry, and steadily improve agricultural comprehensive productive capacity. In accordance with the principles of integration of resources and joint construction and sharing, it is necessary to make overall arrangement for agricultural technology, husbandry and veterinary medicine, animal disease prevention and control, animal health supervision, early warnings in agriculture and husbandry service facilities such as grasslands, agri-economy, agricultural machinery, quality control of agricultural products and agricultural pests, and construction of the agricultural information network platform.

3.8 Taking advantage of local culture and customs through the development of minority's tourism industry in special areas

Many areas of special type are similar to Yushu in Western regions of China. They are the communities of minorities and people with special historical culture and customs, and they have large tourism resources and strong cultural atmosphere. In the past, under the influence of transportational conditions, utilization and exploitation of these cultural resources were restricted. After the completion of the transportation infrastructure of disaster prevention and reduction, development and operation of cultural resources can also become an important element of poverty relief industry. For example, Yushu in Qinghai Province is a minority autonomous prefecture. Tibetans proportion in this area is high. Main Tibetan culture resources are relatively concentrated. These resources include "a district with four zones"and many key scenic areas and spots, such as Jiegu Tourist Area, Tang-Tibet Road Tourism Zone, Plateau Wetland and Grassland Tourism Zone, Kangba Folk Style Tourism Zone, Religious Culture Tourism Zone, Lebagou-Wencheng Princess Temple Scenic, Jiegu Scenic, Batang Hot Springs Scenic, and Lasitong Tibetan Village Scenic, Longbaotan Ecological Tourism Scenic, Saiba Temple Religious Scenic, Sangzhou Temple Scenic, Gongsa Temple Scenic, Sumang Scenic, and Jiatang Grassland Scenic. Futher development and implemention of these cultural resources through construction of transportation facilities not only can disseminate Tibetan culture and customs, and promote cultural exchanges and integration in ethnic regions, but also can explore a road towards sustainable development of poverty reduction in areas of special type.

    To sum up, combining disaster prevention and disaster reduction with poverty reduction is the primary means of implementing poverty alleviation and development in special areas in the new era, and it is also the strategy to adapt to the realities of variable weather and disaster-prone. The key of carrying out poverty work in such areas is to consider and prepare for the frequent occurrence and damage of disasters. It should not only get prepared before the disaster, let impoverished population be aware of harms of disaster through participating disaster prevention and reduction engineering, learning coping measures and training livelihood skills; but also give full consideration into the future changes of poor people's production and livelihood skills and production lifestyle during the reconstruction process after disasters. Only through combining disaster prevention and reduction with poverty reduction, can we better deal with the harms and influences brought by disasters and poverty to poor people in special areas, and achieve win-win outcomes of disaster prevention and poverty relief.

(Translated by Bai Jiawei)

Passion, Ideal and Reality

—the Relationship between One Non-governmental Organization and Rural Communities during the Process of Post-earthquake Reconstructions and Its Significance

Lu Hanwen Yue Yaopeng*

Abstract: During the process of participating into the post-earthquake reconstruction and development in rural communities, some non-governmental organizations (NGOs) are still clearly marked with traditional culture. They incline to rebuild rural communities under their own ideal blueprint, and the relationship of NGOs with villagers and county-level organizations, which formed during the post-earthquake reconstruction and developmental programs, is "authority-obedience". This situation affects the interaction between NGOs and rural communities, actually impacts on developmental programs as well as the growing of civil society.

Key words: Non-governmental Organizations; Rural Community; Interactive Relationship; Development

---

* Lu hanwen, Director of Centre for Poverty and Rural Governance Studies, Huazhong Normal University, Sociology Professor; Yue Yaopeng, Master degree candidate in Sociology School, Huazhong Normal University. Part of this paper has been published in the Proceedings of Guangxi University, 2011(3).

1. Introduction

Chinese government organized and launched grand reconstruction activities after Wenchuan earthquake in which non-governmental organizations played a significant role. ① These civilian organizations had specific characteristics in taking part in post-earthquake reconstruction, like paying attention to proposition of new thoughts, putting field-level community, especially rural villages, as focal points, concerning vulnerable groups, doing work in details, emphasizing ability building and so on. All of these play a complementary role to governmental work.

Environmental Facilitation Association (EFA), founded in 1996, aimed at spreading and practicing ecological civilization. Its main work includes advocating green life, cultivating eco-village, pushing green mass medium. After the May 12th Earthquake, EFA headed for the most serious disaster-stricken areas to investigate and evaluate the disaster situations. Afterwards, EFA hosted relevant forums focusing on ecological civilization and post-earthquake reconstruction, proposed ecological reconstruction and green development. At the same time, it chose Dacun, in PZ city, Sichuan Province, as the experimental spot of practice exploration of ecological reconstruction. ②

Dacun, located on Da Mountain, which is the southern part of the Longmen Mountains, with an altitude of more than 1600 meters above

---

① Zhang Qiang, Yu Xiaomin. *On NGO Participating in Wenchuan Post-earthquake Reconstruction*. Beijing: Peking University Press, 2009, PP. 71-87.

② EFA, Dacun, Dashan ecology association and all people's names are anonym, and did some necessary technically processing. The research aimed at revealing some existing particular problems in NGOs' participating in country development through EFA's program practice in Dacun. The focus of research is the materials implicit in EFA's program practice in Dacun, which have typical significance in understanding the relationship of local organizations and rural community, without involving other activities that EFA has launched these years. Therefore, it should not make comment on EFA, the local organization, according to information in the paper. In addition, there were more interviews on community, relatively less interviews about EFA.

the sea-level, is a village of more than two hundred households. There are fresh air, limpid water, flowing streams and quiet environment. There are also luxuriantly green mountain forests, and herbal medicines which grow all over the mountains. It is of mountain climate which clearly consists of four seasons, and is quite suitable to be summer resort. Meanwhile, local ethos here is honest and simple. Villagers work hard and unite together. Far from urban area, it is seldom affected by the outside world, which makes it indeed a wonderland. However, after the earthquake took place, 90% of houses were ruined and only a few old woodenstructure houses suvived. After field research, EFA formed an ideal of Happy and Harmonious Home("乐和家园") to support the reconstruction in Dacun. In detail, Harmonious Ecology——low-carbon environmental management whose theme is ecological habitation; Harmonious Livelihood——low-carbon economic development whose main body is ecological industries, including original handicraft industry, ecological agriculture and ecological tourism; Harmonious Health Care——low-carbon health care-oriented medical clinics; Harmonious Ethics——low-carbon ethic education whose essence is to respect nature and cherish materials; Harmonious Administration——low-carbon ecological-society organizational system whose speciality is reciprocal and symbiosis. ① Starting from this ideal, EFA gets involved into post-earthquake reconstruction in all sides, and interacts with village-level organizations and villagers.

The article is based on materials obtained from field work, regarding the relationship between EFA and villages as the mainline, analyzing the sinuous story of ecological reconstruction and further discussing some questions like the theories and roles of non-governmental organizations participating in rural developmental work.

---

① Liao Xiaoyi. *Gaze around: Liao Xiaoyi Talking with Sages from Home and Overseas on Environmental-friendly Prescription.* Sanchen Video & Audio Press, 2010, P. 301.

2. The Relationship with "Two Village-Level Committees" (One refers to village-level communist branch committee and the other points to village-level autonomous committee)

2.1 Trust and support: Two village-level committees provide support with great efforts

With the town government's recommendation, EFA came to Dacun to inspect and evaluate disaster-stricken areas after Wenchuan earthquake. The organization held the view that Dacun is endowed with advantageous natural and social conditions which can meet to launch ecological reconstruction, and communicated with village-level communist branch and village autonomous communittee.

> She (who is responsible for EFA) will also have some description of the country. And after taking a look of several villages, she thought this village is suitable. Dacun, relatively speaking, may have less convenient transportation, it has little influence from modernization and the natural environment is protected comparatively well. Although it can be easily seen trashes on the road, it is much better than the surrounding areas of the town when compared with other villages at the foot of the mountains, as it is a place that is the nearest to ideal ecological village. In addition, villagers there are also very active. Therefore we chose this place. "(Interview number: 20101127-02-WP)

"Two village-level committees" welcomed and supported toward the coming of EFA, and was glad to do what ever they can to help EFA to carry out ecological reconstruction — building Happy and Harmonious Home.

> At the beginning, "two village-level committees" was very thankful to EFA participating into post-earthquake reconstruction, and we also feel grateful to them. Both sides carry a grateful heart inside. (Interview Number: 20101129-02-LG)

EFA chose the whole land of villagers of group eight, nine, ten and eleven which were situated in higher altitude in Dacun, and some and of group five to build Harmonious Home. Before EFA came, government provided five alternatives to reconstruct houses for those villagers: unified plan and unified construction, unified plan and self construction, self construction on former place, reconstruction on former place and money arrangement. Due to the inconvenient transportation from Dacun to TJ town, most villagers in Harmonious Home community plan to choose unified plan and unified construction or unified plan and self construction to move from present living places on mountains, where generations had lived, to settle at the foot of mountains—TJ town. However, "Harmonious Home" program would rather villagers continue living on the mountains. From this point, it needs to mobilize villagers who have chosen unified plan and unified construction or unified plan and self construction to transfer to choose reconstruction in former places. "Two village-level committees" had done a lot of work to help EFA to persuade villagers to accept reconstruction in former places.

> Chairman L, came here and made us choose reconstruction in former places, thus can this program be settled down. At that time, Secretary LFG and chairman L had been spreading the advantages of reconstruction in former places, and they also organized and mobilize villagers to start about the reconstruction. (Interview Number: 20101128-02-XSM)
> Ask: When did EFA come here?
> Answer: the situation after the earthquake was like this, EFA first developed in group ten and eleven. All of us didn't join in until August. There were "double Ls" in that period, LFG was the secretary of villages and had persuaded us more than one month. We signed to unified construction at first while changed to choose reconstruction in former places just after accepting their persuasion.
> Ask: as we know, EFA didn't choose group ten and eleven, neither

your group, but how do you join into "Harmonious Home" ?

Answer: In fact, all of us were not willing to join in it, but some people came here to persuade us and claimed that their main objective was to develop tourism and rural entertainments, by holding meetings and repeated persuasions, with a sense of hardness and enforcement. It takes one month for LFG to transfer from the town to here. (Interview Number: 20101127-01-HJW)

After the program starting up, "two village-level committees" gave EFA great support and coordinated with it to establish Harmonious Home management community, which was proposed by EFA, village autonomous communitee and Da Mountain Ecological Association, when EFA proposed a negotiation mechanism of Harmonious Home. The three parties signed "Dacun Harmonious Home Cooperation Agreement", made definite tripartite responsibilities, and confirmed each other's rights and obligations, which laid the organizational and institutional foundations for implementation and management.

In the initial stages of reconstruction, funds and other resources brought by EFA and the passionate construction idea of Happy and Harmonious Home won the trust of "two village-level committees". Meanwhile, this resulted in "two village-level committees" gave great support to the preliminary work of EFA.

2.2 Disappointment and repulsion: "Two village-level committees" bifurcated

With the advance of Harmonious Home program, EFA and "two village-level committees" have divergence in program planning and management and development path of industries.

During the planning of Harmonious Home program, EFA put forward basic such building content as "cultivating, reading, touring, handicraft and regimen", distinguished the land into cultivated land from non-cultivated land which should be used according to the principle of "unified plan, centralized control and production in each distributed household". Village-level autonomous committee viewed what EFA proposed as an "ideal plan", a portrayal concerning ultimate blueprint, without bringing forward operable steps and concrete measures about how to realize this blueprint. Thus they ngotiated with EFA several times on this point, advised EFA to set forth an operable implementation plan which corresponds to actual developmental needs. However, every negotiation failed to reach the anticipated effect.

What chairman L mentioned to us was Harmonious Home, I has been always requesting L that whether she can offer a detailed program, no matter how far away the program it was, while we can put it into practice in several steps, like what goal we can realize and what can be made to come true next year. But what L chairman showed to me was only her conceptual plan. What is called conceptual plan is without detailed substance in the program. Instead, it is an ideal thing. "cultivating, reading, touring, handicraft and regimen", which was the word that She often talked, and what she portrayed was the prospect of ultimate phase. In fact, it has many steps that needed to accomplish from present period to ultimate phase, but she had no scheme about how to fulfill the whole process. For instance, in my opinion, it needs normative implementation project, and EFA help us to carry out post-earthquake reconstruction. To start with, they should make us known that their financial capacity and how much fund they have, and how much fund can be used in program management, and how much could be used as the input. Supposing that two million or one million will be invested into construction, what should we do this year and next year; what goal can be reached, what achievement can be seen after the fund investment, all of which need sufficient communication with local authorities. I had communicated with chairman L several times, but communications failed. Thus I don't want to communicate with her any more, just because I made it clearly every time, but my anticipated result failed. (Interview Number: 20101129-02-LG)

Consequently, EFA didn't accept the suggestion of "two village-

Passion, Ideal and Reality

level committees", instead it continued to push Harmonious Home program following "ideal plan". Volunteers and servicemen made up the EFA program management team. According to "Dacun Harmonious Home Cooperation Agreement", the team, Dashan Ecological Association and "two village-level committees" make up Harmonious Home management committee which is responsible for making strategic decisions about significant matters in the process of program implementation and supervision through holding tripartite joint meeting. However, "two village-level committees" view that tripartite joint meeting mechanism is not executed as stipulated in the agreement.

What EFA lack is custody and supervision in the operational process. Formally, we established an institution which is called "Tripartite Joint", but it only exists in name. Why speak like that? Firstly, quality of common people do not meet the request of management and their quality are quite low; secondly, financial affairs are never open to the public. I still don't know how much fund EFA had put into Dacun, and how much budget in the prophasic work used to input the fund; in addition, the group who take charge of program implementation is not a very professional team, it is a group that is composed of many volunteers, and those volunteers are not specialized in techniques and management, not expert in industry development.

NGO is just a platform, but it is not omnipotent, neither is equipped with capacity of implementation the program. Or it should have an implementation body of the program, which should have implementation scheme and capacity of program supervision, while all of these do not exist. How can our field-level institutions communicate with NGO equally? We have no idea about how to communicate with them. Maybe in the eye of NGO, our quality is too low, we cannot understand their ideal, and we may appropriate or abuse the funds they give us. Of course, it may form team, but the team they set up is not a team that a lot of volunteers take part in. Instead it indeed is a team with implementation steps. All aspects should participate in it and finance should be transparent. As a NGO, the finance is not open and clear, it will lose the trust from local authorities at the beginning. (Interview Number: 20101129-02-LG)

Hence one can see that, the discontentment of "two village-level committees" toward tripartite joint meeting mechanism mainly concentrate on two aspects: one is that they hold a negative attitude toward management quality of ecological association; the other aspect is that they doubt and distrust professional capacities and financial management work of program implementation team of EFA.

Due to the lack of effective communication between the two parts, the divergence in "two village-level committees" and EFA expanded further to industries development and projects shares of stock and allocation of bonus.

In the industry development path, EFA established organic farm in Harmonious Home, popularized the plant of growing crops, like organic vegetables, kiwi fruit and Chinese yam to local villagers, running in the mode of "public welfare and farmers", produced in households, unified selling, at the same time, promised to guarantee the selling channels. Nevertheless, in the view of "two village-level committees", at first, villagers have the right to choose whatever to plant freely and any organization has no right to ask villagers to plant the kind of crops; secondly, local climate environment is not suitable to popularize the crops that EFA proposed; once more, EFA failed to fulfill the promise of guaranteeing the selling channels. The mode of "public welfare and farmers", which EFA had advocated, didn't help villagers realize increasing incomes, on the contrary, it restricted the development of local economy. What's more, it transferred the negative influence brought by the program to "two village-level committees".

> We have no right to decide on what to plant in the villagers' land. What we can do is only to suggest. To suggest farmers plant what crops or do not plant what crops. However, some handling ways of NGO

requested villagers must plant what and ask us to disseminate to local residents. I said that I cannot do in that way. If it can get some earnings, it would be ok; but if it turns out to have no incomes, who will take the responsibility? Since the land has contracted to villagers, then they have the right of running and management. We can only conduct and analyze in techniques, like what is fit to plant on local land, what the trend of market will be like next year. In terms of this aspect, we had a large contradiction with NGO on this affair. They considered our village-level committee do not support them, and misunderstood that as the secretary of communist branch, I have the right to determine villagers to plant kiwi fruit and Chinese yam. Thus, I think NGO is not familiar with field-level organizations.

Chairman L introduced some industries, like kiwi fruit and Chinese yam, but they are not suitable to plant in the local land. The situation of test-plant is not ideal and satisfactory. Therefore, it hit the positivity of local villagers in this aspect. The reason why it was unsatisfactory is that they failed to combine it with our local real industry situation. Since we established the Harmonious Home in our place, it needs to be integrated with the local climate, soil and environment. (Interview Number: 20101129-02-LG)

NGO intervening into the post-earthquake reconstruction not only failed to boost the local development, on the contrary, it restricted local growth. It has been two years. In fact, it transferred the pressure of failure to our villages, and resentment and discontent of villagers have been shifted to the two village level committees. Because they are not local people, thus local field-level organizations had no choice but to undertake the negative influence they brought about. In addition, it communicated little with "two village-level committees" and local authorities. They work just in their own way without communicating with us. Whether it is fit for local situation or not, they neither seek for our opinions. As a result, it led to some unnecessary effects on local residents. As long as we put forward some suggestions to them, it seemed that we do not cooperate with their work. Moreover, they asked us to do some dissemination. While if we propagandize and disseminate their ideas to the public, the bad influence will become more far-reaching. (Interview Number: 20101129-03-LY)

As for the shares of stock and allocation of bonus of Harmonious Home program industries, EFA becomes a shareholder with funds and ideal, and villagers buy a share by land and some other natural resources. EFA takes up 51% of the stock and bonus, and the Ecological Association holds residual 49%, then distributed the shares to villagers in the form of bonus. "Two village-level committees" queried the way of allocation, and considered that the allocation way of shares and bonus is lack of basis and thus is unequal, and the ideal of EFA is not matched with the reality of villages.

> "The most maximal disagreement is that chairman L established tripartite joint meeting, drafted the agreement, and there is a statement about the allocation of share and how to distribute when having earnings. The side of chairman L holds 51%, and local villagers possess 49%. Well, she mentioned that EFA takes up 51%, taking out as the expenses of exploiting the market, cost of operation, etc. Then the rest would be allocated to villagers. It sounds good." However, I noticed the question in it, namely, what is the standard of the share allocation? Chairman L wants to become a shareholder with funds, we have local resources, so it is reasonable that we hold 49%. If you only take with an ideal and assumption, and you want to possess 51%, it is not fair. Although Chairman L won't take the 51% away, what in her mind is a rolling development, the local villagers cannot figure out why is allocated in this way. In my point of view, the reason why the ideal is not in line with the reality is that there is a long way to improve the quality of villagers. To endorse her ideal may realize the real Harmonious Home if we work arduously for one decade or two decades. After all, I still claim the same words: The long march is great, however if numerous people died of

starvation on the road, what is the meaning of long march then? (Interview Number: 20101129-02-LG)

On account of above divergences, "two village-level committees" communicated with EFA for several times, but none of them reached a concordant opinion. Considering EFA is the inverter, and out of respect for person in change of EFA, "two village-level committees" did not insist on tackling the problems according to their own position. However, the villagers lost the initative to cooperate, and gradually became disappointed and repulsive to the Harmonious Home program, then they bifurcated and went back to their own ways.

> When it comes to cooperation, how we cooperate with them is just to let Ecological Association and us join Harmonious Home management committee, while all the other cooperations were undertaken on their own. It was only when building houses that Ecological association participated in.
> EFA played the leading role, and what role "two village-level committees" played was just on the surface. That means, we seldom took part in it. Honestly speaking, local villagers didn't know what they were doing. How is it possible that we know it very well, let alone the villagers. You just mentioned about tripartite joint meeting or a kind of cooperation. But what is cooperation? Cooperation should be based on the foundation of mutual trust. Since "two village-level committees" stand for the interest of common people, and Harmonious Home represent the interest of EFA. Thus There is contradiction between bilateral interests. We have no idea, but we must scheme villagers' existence and development in the future. There exists a big problem on cooperation. To tell the truth, I have come here for more than one year, but we have never worked together with them.
> Our "two village-level committees" would not give up the villagers who reconstruct in the former places on the mountains. No matter the government nor "two village-level committees" will abandon those one

hundred households, because they are villagers of Dacun, TJ town. Even though the process will be extraordinarily tough, there will be nothing to be feared as long as local villagers stand together with us. Therefore EFA has to transform the way of dealing affairs. (Interview Number: 20101129-01-WPY)

3. Relationship with Village-level Ecology Association

3.1 Conduct and cooperation: The birth of ecology association

To fully mobilize villagers' initiative to take part in construction of Harmonious Home and establish effective villager-participation mechanism, EFA lead residents of Harmonious Home program community to set up "Dashan Ecology Association". It was the first civil environment-protection orgnization. Dashan Ecology Association sets positions like chairman, vice-chairman, chief secretary and accounting custody, draws up its own regulations, and is enrolled and registered in the Civil Affair Departments. Elites in the community undertook the person in charge and manager of Ecology Association. At the same time, they were also the earliest villagers who received the ecological reconstruction ideal of EFA.

In the early period of foundation, Dashan Ecology Association viewed that construction ideal of Harmonious Home was very advanced. If it can get the aid of influencial power of EFA, the livelihood program whose core is ecological tourism will be able to develop, and villagers' increasing incomes and becoming rich can be realized soon. Based on this understanding, Ecology Association positively assists EFA in explaining the content of programs and ways to improve incomes. Villagers gradually accept building goal of Harmonious Home and submitt applications to become members of Ecology Association. What is worth mentioning is that dissemination and explanation of Ecology Association is an important reason that villagers gave up unified plan and unified construction or unified plan and self-construction and turned to reconstruction in former places.

"While it does not mean that it is communist secretary made the decision. But what is the reasons of this? Chairman L said to develop ecologial tourism, so we chose this kind at that time. What people considered was that chairman L wanted to develop ecological tourism, and to our individual, it doesn't matter to choose unified construction or others. People thought that since chairman L claimed to develop ecology tourism, thus we choose reconstruction in former places. Well, you know, local villagers would not think too much or consider too much, and just made the choice immediately. Generally speaking, people considered that chairman L claimed to develop ecological tourism, all people just put all minds on this." (Interview Number: 20101112-01-XEQ)

The first concrete work of cooperation between EFA and Ecology Association was housing reconstruction. EFA applied for housing reconstruction aiding funds from large foundations. After application succeeded, EFA remitted the funds into the account of Ecology Association which is granted to Harmonious Home community residents ecology association. Then, EFA invited engineers to design houses for Harmonious Home community residents. After the blueprint is finished, Ecology Association invited some tectonic-experienced villagers to discuss and alter the blueprint with engineers, and finally to figure out villagers' favorite building style. During the process of reconstruction, EFA took the advantages of external funds and other resources, and Ecology Association provided local experiences and human resources. Both sides achieved mutual complement of advantages.

3.2 Control and affiliation: The metamorphosis of ecology association

Like the "two village-level committees", some divergences emerged between EFA and Ecology Association in the process of Harmonious Home construction. They mainly exhibited in three aspects: industry developmental paths, financial custody and bilateral role orientation.

On industrial developmental paths, although both sides agreed with Harmonious Home building ideals and developing ecological industries, two parties displayed different opinions on developing strategy. EFA started up ecologyical tourism, organic agriculture and original handicraft industry in the concrete implementation process. Ecology Association figured that funds should be focused on the construction of tourism supporting facilities at first and develop organic agriculture and original handicraft industry after ecological tourism had developed. According to viewpoints of a person responsible for Ecology Association, what villagers mostly want to see is that programs can actually bring about improvement of incomes. It would cost greatly to develop organic agriculture and original handicraft industry when ecological tourism still did not develope. It perhaps cannot promote the increase of incomes, instead may lead to large funds deficit. Meanwhile, he mentioned that to develop rural entertainment in the approach of "centralized management and unified allocation", which EFA put forward, was unfavorable to integrate funds and mobilize villagers' participating initiative.

"I believe it is feasible to develop tourism in our place, what needs is to spare no efforts to make all constructed housed into scales and then to develop rural entertainment. It calls for further investments. But after the earthquake, local residents spent the money on building houses and furnishings. In most cases, it cost more than ten thousand yuan, which is really a large loss. Such situation did not allow each household develop rural entertainment by themselves but should be unified. In my opinion, if every family can start singly, they can develop it faster by loan. However, EFA adopted the way of centralized management and unified allocation, thus you had no choice." (Interview Number: 20101127-01-HJW)

In the aspect of financial custody, EFA managed all the other program funds except handing over house-reconstruction funds to Ecology Association and then granted to villagers. During the process of program implementation, EFA didn't set up financial supervision group with Ecology Association as the stipulation in the contract to custody finance together. After the program ended, EFA also did not open the program funds using state to Ecology Association. Ecology Association regarded they had the right to supervise the finance, and requested EFA to open finance to the public several times, but failed to get any reply. As a result, Ecology Association started to suspect EFA had some financial problems in the process of program implementation.

> When the time I left, since it was a "tripartite joint meeting" institution, then finance must be open and transparent. All program investments should be overt to guarantee people know where and how the money was used. Now that we cooperated, villagers have the right to know the internal facts. Let villagers know utilization of each budget. All things were agreed on how to put into practice when I left, while none of them has been turned into actions. Up to now, many affairs are still in a mess. (Interview Number: 20101128-01-YGL)

> Ask: does Ecology Association have its own account and custody?
> Answer: yes, it has. But there is nothing in hands. It is blank and only in form, without real meaning. There is no necessity to talk about whether it has or not.
> Ask: normally, what role can account of Ecology Association play?
> Answer: Ecology Association should have the right to know the internal information about all finances between EFA and Ecology Association and all funds of Harmonious Home. So does the right of funds-using and allocation. Now they are utterly ignorant, and know nothing. (Interview Number: 20101128-01-YGL)

Concerning the orientation of bilateral roles, Ecology Association views EFA led the plan, decision and implementation and custody of the program, while Ecology Association mainly played the role of supporting EFA to implement the program. Although Harmonious Home management committee once held tripartite joint meeting, suggestions that Ecology Association put forward were often not adopted, so Ecology Association considered it was only an empty form, without exercising rights that the tripartite agreement stated.

A leader of Ecology Association showed that they had spent a lot of energy on Harmonious Home, which even lead to the decline of family incomes. But EFA didn't give any relevant payment and compensation, which is inappropriate.

Since EFA mastered funds and relavant resources, controlled the voice right. Ecology Association lied in disadvantageous condition when having conflicts with EFA. Thus some leaders who have inconsistent opinions with EFA resigned.

> Well, I just had witnessed something, and pointed out which made some of you felt uncomfortable. You claimed that if I had any problems, just call you, but when I ring you, you seldomly listen to me. Well, you seldomly listen to me, why should I call you again? I mentioned SG, the person that you trust most, you just said it was wrong that I pointed out her flaws. What can I do, how can I work for you? (Interview Number: 20101128-02-XSM)

EFA turned to support those leaders who were more willing to cooperate with them. And the relationship between EFA and Ecology Association metamorphosed into control and affiliation.

4. Relationship with Villagers

4.1 Gratitude and response: Villagers' first choice

At the time when Post-earthquake reconstruction just started up, some Dacun households who became members of Harmonious Home community later have the inclination to choose unified plan and unified construction or unified plan and self construction, thus they can grasp this opportunity to change living places to settle at the foot of mountains.

> In the past, our village was very poor. Villagers were not willing to

build houses on mountains for the sake of next generation. Just hope they can be better. Thus most of us choose this "unified plan and unified construction". (Interview Number: 20101125-01-YBR)

After the earthquake happened, local people want to move to live at the foot of mountains and settle home in TJ town. While we also have some worries, namely, every family has more than ten acres of Chinese goldthread on average. If we moved to the foot of mountains, we have to walk more than five kilometers footpath to the top of mountains, which is inconvenient. Government carried out five reconstruction policies: unified plan and unified construction, unified plan and self construction, self construction on former place, reconstruction on former place and arrangement with money. What we choose was unified plan and self construction. (Interview Number: 20101128-02-XSM)

It was villagers' worries of moving to the foot of mountains that made EFA see the chance of persuading villagers to choose "reconstruction in former place" to carry out Harmonious Home program. And then EFA pushed to hold villagers meeting to analyse advantages and drawbacks of moving to other places.

Perhaps because government carried out five choices of moving method, and villagers could choose one of them, all local people chose "unified plan and unified construction". Consequently, EFA held meetings everyday, or even several meetings in a day to persuade people to remain.

EFA said if you choose unified plan and unified construction, what do you live on after moving to the foot of mountains? And I answered, we can grow our land at home, to plant crop in our own lands, it is best that each family has some amount of land. Some talented people claimed to do some manual works in the cities. EFA then said, after the earthquake, many people were out of employment, if you go out to seek for a job, how can so many people find a job? If staying on the mountains, people in their sixties and seventies or fifties, female or male, young or old, all can find employment in Harmonious Home. (Interview Number: 20101125-01-YBR)

Obstructions which comes from worries of inconvenient traffic on the mountains and difficulties of finding employment in cities to maintain the livelihoods and attractions of EFA' Harmonious Home program which promised to improve incomes made villagers begin to weigh advantages and disadvantages of reconstruction in former places and "unified plan and unified construction or unified plan and self construction".

Villagers reflected that it was inconvenient to go outside if staying on the mountains. Then EFA made commitments to villagers that it would negotiate with the government to construct a mainline road in Dacun. With the promotion compaign of EFA, government of PZ city quickly contributed investment to constructing the road and the completion of the road in Dacun makes it convenient for villagers to go outside. Villagers witnessed EFA could fulfill its promise of negotiating with government and got to know the social influences of the institution and people in charge via mass media. Thus they became surer that EFA has the capacity to bring in funds and relevant resources to help them increase incomes and become rich. Then they decided to choose reconstruction in former places, and stay on Da Mountain to build the Harmonious Home with EFA.

> My wife is native of Hebei province and planed to leave and never come back. There was no interest to live on mountains. Paths on mountains are narrow, full of mud and water. We can't carry anything to the foot of mountains. The key was the road. If there was no cement road, it would be better to live down the hill. Well, with the cement road, our situation is indeed different from before. It makes no difference between the top and the foot of mountains. In addition, the other advantage of living at the top of mountains is that if farmers want to get something, just walk out of the door without going too far. While at the

foot of the mountains, if you have no job or if you are not working for a big corporation, you may have no source of income. The overall development of above-mountains tends to ecological tourism. Well, to think what you can get if you live down the hill, basically, nothing can be obtained. With the increasing of age, most of these people are in their forties or fifties, so usually there is no work they can do. Well, most of them would be very unhappy. If property management fee and other kind of fees are charged, they would be unhappier. At least, there's no need to worry about these fees on the top of the mountains. (Interview Number: 20101128-01-YGL)

Well, let's talk about chairman L. Why does she come here to do this job (Harmonious Home program) in such an old age? You also watched TV, she worked for us to construct the road. In other words, what does she work for? We live on our own, and life will become better and more comfortable. So please be cooperative. (Interview Number: 20101125-01-YBR)

At this period, villagers were quite grateful to EFA and actively responded to the appeal of building Harmonious Home.

4.2 Indifference and alienation: Villagers' second choice

In the early stage of Harmonious Home program, villagers supported and responded to EFA's work with great enthusiasm, filled with confidence and hope about future development. While with the implementation of the program, the points that villagers dissatisfied with became more and more.

Harmonious livelihood contains original handicraft industry, organic agriculture and ecology tourism. Original handicraft industry refers to embroidering handkerchief. According to the thought of EFA, this kind of embroidering handkerchief not only embraces humanistic values, but also is reusable, equipped with the characteristics of low-carbon, thus it has markets. However, the real situation was, EFA started up this work and paid women who took part in embroidery wages by the piece of work. And their daily pay was approximately from

20 to 40 yuan. Planting and processing of local leading industries—Chinese goldthread belongs to labor-intensive industry, and it also has necessity of employment when it's in busy season. And general pay was 50 yuan a day. Additionally, free melas was provided. As a result, women had little motivation to embroider handkerchief because of low level of wages, especially when the farm work is heavy and busy.

> Well, you embroidered the handkerchief, and just get the pay of 30 yuan one day, without free meals. Later, we didn't go there. You see, if you help villagers to dig Chinese goldthread, you will receive 50 yuan a day with the three meals a day. Nowadays, price of commodities rise, the pay of the work was too low. If you plant Chinese goldthread, which can be digged for three to five years, you can earn money. Thus people prefer to grow Chinese goldthread and do farm work. (Interview Number: 20101123-03-LGR)

What's more importantly, although embroidering handkerchief can sell at a high price in theory, and have market potential. However, due to the fact that ecological tourism still didn't grow up, marketing work lacks factual performances. Except for sending original handkerchief embroidered by women in earthquake-stricken area to celebrities, like Clinton and Bai Yansong, which achieved certain propaganda effect, original handicraft industry of Harmonious Home failed to achieve market benefits. Handkerchief processing workshops came to a standstill because of no profits after running for a period of time.

> They set up workshops and cost a lot of money to buy equipments and materials. Meanwhile, they mobilized our males to join into the process of handkerchief production. Do you think this is realistic? Moreover, insiders claimed that embroidering products still keep long in stock in headquarters storage. (Interview Number: 20101129-03-LY)

Ecological agriculture refers to building organic farms and planting organic crops. The development of this industry involves three conditions: The first one is enough farmyard manures to reach the effect of replacing chemical fertilizers; the second one is sufficient and low-price labors to clear weeds and catch pests by manual work thus to realize the impact of substituting for weedicide and agricultural pesticide; the third one is organic products can be sold at a high price thus it will not only cover the cost of labor input but still have space to make profits. However, practice indicated that farmyard manures cannot ensure high outcome, harvests of organic agriculture obviously declined. The prices of organic products market can be relatively higher, but when the price is too high, the demand of market decreases.

> Ask: how about the sale of hogs?
> Answer: it depends on market.
> Ask: do you sell the hogs to the market directly?
> Answer: yes, according to the market price, we want to sell the hogs at one price, But EFA said to sell at another price without fully considering whether customers can afford it. So its claim was unrealistic. I started to feed hogs on forages and in later period used the foodstuff, to save cost as much as possible. We need to provide hogs that should be affordable for customers. (Interview Number: 20101124-01-HJF)

Wages that organic farms of Harmonious Home were willing to pay cannot arouse the villagers' activeness to engage in the work. Thus it cannot normally run. Some land was left unused and at the state of desolation. Others which had planted crops in fact were at the state of emerging and perishing by themselves.

> Last year, if chairman L said something, we all did our best and spared no effort to manage it. Yet now no matter how many fine words she said, all would fail to embrace hearts of villagers. Above institutions hired us to dig Chinese goldthreads, 50 yuan a day including meals. While doing farm work we can only get 30 yuan a day without meals, so nobody

will be willing to work in Harmonious Home. (Interview Number: 20101125-01-YBR)

Ecological tourism refers to developing the leisure-life and regimen tourism relying on local natural ecology. But it faced lots of problems: short of funds, infrastructure and supporting facilities were poor. Apart from the fact that EFA solicited a few businesses of trainings and inspections, ecology industry of Harmonious Home virtually became half-working project.

> Now people began to worry. Although Harmonious Home cannot be built up in a day, it needs to start, and then gradually develop better. We also want to develop tourism, but to tour what? What to see and play? For instance, some outside people come here, they must consume, so food drinking, and entertaining should be upgraded and some flowers and plants should be furnished to make houses more beautiful. Well, chairman L just claimed there was no money. But if no money how can she build houses (public study building and farmers' chamber)? What do these for? Why did he spend nearly one million to build these? Instead, it should build some practical and worthy ones. For example, it's quite hot in summer. She claimed to promote consumption, but tourists must come here just for joy, for fun. In fact, no entertainment places can be found here. Many people had given suggestions to chairman L, while EFA replied there was no money. Our villagers only can rely on ourselves without any other choice. Depending on limited and scarce support from EFA cannot meet the need to survive. Everyone just got 66 yuan this year, what can these do? It would be ok if there weren't many children in the family. If the family have a child at school he will come back home every Friday then leave every Sunday, thus it calls for basic living expenses. Besides fares, one child at least need 20 yuan. If we cannot find 20 yuan, how could we supply our children? (Interview Number: 20101125-01-YBR)

Harmonious ecology mainly centered on houses reconstruction.

EFA granted each villager a subsidy of 4,850 yuan per person for housebuilding. Although it was less than subsidy standard of unified plan and self construction which was 8,000 yuan per person, villagers had no obvious disagreements or objections. In their view, if the tourism industry can develope and blossom, corresponding incomes sources were more important than lump-sum building subsidy. But just as mentioned above, tourism industry failed to achieve anticipated effect as what EFA had portrayed. Villagers began to complain, even regret to stay on the mountains to reconstruct in the former places.

> We has lived on the mountains from generations to generations. The economy is under-developed, and conditions are tough. Daughter or son, all people moved down the hill, only left our aged people on mountains. We always worry the things in the future. For instance, without source of income, without money, how to take a taxi even when roads have been constructed. If Harmonious Home succeedes, it can promote the development of economy. All people will be content, supportive, and grateful, and they will do their best to push foward the work. However, if Harmonious Home really cannot develop and thrive, all villagers will be disappointed. (Interview Number: 20101125-01-YBR)

We can draw the conclusion from what has been discussed, although the sheer woody houses, which are based on traditional civilian housing style and were designed and constructed after scientific improvement, they have the advantages of nature, beauty, firm and environment-friendly, which really highlighted the significance of ecological reconstruction. But villagers seemingly cannot figure out the advantages they have possessed. The new houses failed to retain villagers' minds and sights. On the contrary, the infinite possibilities of generating incomes and improving livelihoods from the external world engendered their endless longings.

Harmonious health care and harmonious ethics are two aspects that villagers feel remotest in Harmonious Home program. Harmonious health care involve country clinics, whose root is Chinese traditional medicine culture and regimen sports. Clinics broke down because it could not find doctors who were willing to work there, and regimen sports could not attract the interest of villagers.

"Since it was health station, and four groups totally have 280 to 290 people, you must hire a professional doctor here. Salary, treatment, medical insurance, endowment insurance should keep pace with the normal standard. All of these would cost a large amount of money. Therefore, it cannot afford to hire a professional doctor. Of course, this was a good idea. However, rural entertainment and tourism are still under-developed, because of lack of environment resources. (Interview Number: 20101127-01-HJW)

As for health care:

Ask: now it promoted regimen sports in village, did villagers take an interest in it?

Answer: it seems that villagers were not interested in it. It was ok when people played together in the past, it became worse when people do it on their own. (Interview Number: 20101112-02-LLS)

Harmonious ethics advocates Chinese traditional values like the harmony between human and nature, etc. , but these were not rooted in villagers. Some villagers reflected that EFA supported people who had the consistent opinions with them in Ecology Association and process of other public programs construction and management. Thus the relationship among villagers became less united and harmonious than the time before the program was implemented.

When villagers gradually engendered complaints against Harmonious Home and thought they had lost developmental space because of staying on the mountains, EFA didn't give effective replies, but turned to complain villagers didn't know how to be grateful.

In recent years, most people don't believe her anymore. Previous promises are not delivered. Thus no people would believe her. We proposed some suggestions to her, but she said we were not grateful and discontented. After she made such claim, we no longer want to speak with her. (Interview number: 20101123-03-LGR)

Under such background, villagers began to find approaches to improve incomes and gradually lost interest in Harmonious Home program and tended to alienate and be indifferent to it.

It was when we brought back 66 yuan from her while other people had finished unified plan and unified construction, we lost the trust on EFA. Well, you see, how can we live on so little money? We can not just depend on EFA. What can we do is only depending on ourselves. (Interview Number: 20101125-01-YBR)

5. Summary and Discussion

5.1 Idealism and realism: Choice of village reconstruction and developmental path

Global warming, environmental pollution and ecological degeneration have become the common challenges that the globe faces. EFA declared and worked for coping with these challenges since it was founded, and formed the ideal of environmental protection by using oriental wisdom to cure the ills of western civilization[①]. Under the background of large-scale reconstruction after Wenchuan Earthquake, judging from the point of view of coping with the environmental challenges. Undoubtedly, it was correct and far-sighted that EFA put forward the ideal ecological reconstruction. However, the practice of ecological reconstruction in Dacun Harmonious Home not only failed to be firmly rooted, but also it

---

① Liao Xiaoyi. *Gaze around : Liao Xiaoyi Talking With Sages from Home and Overseas on Environmental-friendly Prescription*. Sanchen Video & Audio Press. 2010, P293-300.

became more and more obvious to see the trend of alienation and abandonment in villages.

EFA was just like an ideal organization in Dacun. It set up an ambitious and grand goal, portrayed a living blueprint of harmonious relationship between human and nature, merging between heart and external objects, assumed this was the world that everyone wanted to have. Villagers and organizations of Dacun were realistic. They accepted ideal of ecology reconstruction. However, what they want first are to recover and improve the present livelihood, earning more incomes and living a richer life.

The problem was whether human can harmoniously get along with nature? Especially, when a large range of countries and international society are still in great tension between human and nature and quarrel ceaselessly because of climate changes, one village which was underdeveloped but had been deeply involved in modern market system, can it be the first to abandon mainstream developing model in modern world and turned to lead the world? Can this village create and support one huge and solid conceptual system and resist and rebuild the mainstream world? Provided answers of these questions are not so certain, then it is not strange that EFA encountered predicament in its practice of ecology reconstruction in Dacun.

The further problem lies in modern society with distinct differentiation: Is it valid for villagers who live at the bottom of society to take precedence to chase for improving livelihood and getting rid of poverty and becoming rich? Supposing the answers are affirmative, we cannot use short-sighted and use other similar words to comment on or criticize these villagers and organizations. After all, center cities which have more obvious "modern problems" only achieved limited success in the pursuit of low-carbon life. Therefore, considering EFA confronted dilemma in Dacun, the responsibility lay in itself instead of villages. In fact, like EFA, Dacun also hopes not only to recover and enhance livelihood conditions, but also step onto a low-carbon development road. But the difference was that when livelihood became "a castle in the air", it had little possibility for villagers to still persist in "Harmonious Home", while EFA was still willing to continue to defend its ideal.

Undoubtedly, Western civilization calls for criticism and reconsideration since Industrial Revolution. Chinese rural development is necessary to obtain some wisdom from this criticism and reconsideration. However, on the contrary, to simply imagine traditional country that lags behind can easily step on an ecological developmental road, which unavoidably carries some Utopian characteristics. Realism may lead to be short-sighted, while separation from reality will get us into trouble. Lacking ideal, human beings will run short of passion and intelligence. But idealism sometimes brings about tragedy to the world. Country reconstruction in disaster areas and even the whole country development in China cannot lose ideal, but at first it needs to base itself on actual demands of farmers.

5.2 Redeemer or partners: Roles and effect of NGO participating in village reconstruction

China is a country with deep idealism traditions, and the ideals of "internal sage, external king" and "Great Harmony Society" have continued for thousands of years without break or abandonment. Around 1920s and 1930s, a large number of people with ambitious lofty ideals embraced this ideal and walked along the road of saving the countryside and salvaging China. Transformation movement of the countryside stepped up on the historical stage in various concrete configuration and shapes. As eighty years have passed, there still are a number of elites in society to live and declare for remolding countries in China. Well, a lot of differences have taken place during the eighty years, but one point remains unchanged, that is the "sage ideal" and complex of salvaging the world[1].

---

[1] Lu Hanwen. *Conferring Rights and Innovation of Poverty Alleviation Institution*. Asia development bank research report, 2009, P. 1-5.

Some people said, the year of 2008 was the first year of civil society and volunteer actions in China, and Wenchuan earthquake ended the debate about whether it has civil society in China or not[1]. But is it the real case? The author views that, some conditions for the growth of Chinese civil society are still at pregnancy and breeding. But many volunteer actions emerged after Wenchuan earthquake and post-earthquake reconstruction still carried striking traditional marks. Some NGOs also rely on charismatic authorities to obtain resources and launch activities, which is the continuation of traditional "internal sage, external king" ideal. EFA's practice of ecological reconstruction in Dacun demonstrated this point.

When dealing with the relationship with "two village-level committees", Ecology Association and villagers, EFA was more like an external authority. It propagated its reconstruction claim and blueprint, tried to persuade villagers to believe these claims and blueprint were "good", mobilized and even requested villagers to support it to launch activities. This was not fellowship, but the relationship of authority and obedience. This unequal relationship didn't arouse resistance from villages in the beginning, but when the wills and rights of villagers and organizations failed to get effective reply and consideration, problems emerged. The ecological reconstruction program in Harmonious Home became the monodrama of EFA, and "external authority" lost recognition of community.

In a well-developed civil society, civil rights and responsibilities are the foundation of social relationships and order[2]. NGOs in development fields are the promoter rather than the redeemer, as well as the partner to share pains caused by the development. Although "participating development", "conferring-right poverty relief" have been regarded as

---

[1] Xiao Yanzhong. *To Develop under Disasters: Wenchuan Earthquake Witnesses the Growth of Civil Society in China*. Beijing: Peking University Press, 2009, Preface and page 10.

[2] Turner. Civil Identification and Social Theory. Jilin: Jilin Publishing Group Co. Ltd, 2007, PP. 163-186.

basic ideals of post-earthquake reconstructions after Wenchuan earthquake and the whole country developing practices in China. However, even those civil organizations which highly advocate flags of "participation" and "conferring-right" sometimes appear as the role of external authority in villages. When villagers can indeed get the actual benefits, they will be glad to accept this "external authority". Logics of Chinese traditional culture instead of that of civil society plays greater role in it.

References:

[1] Zhang Qiang, Yu Xiaomin. On NGO Participating in Wenchuan Post-earthquake Reconstruction [M]. Beijing: Peking University Press, 2009.

[2] Liao Xiaoyi. Gaze around: Liao Xiaoyi Talking with Sages from Home and Overseas on Environmental-friendly Prescription [M]. Beijing: Sanchen Video & Audio Press, 2010.

[3] Lu Hanwen. Empowerment and System Innovation of Poverty Reduction [C]. Report of Asian Development Bank.

[4] Xiao Yanzhong. Challenges Make a Nation Much Stronger: Wenchuan Earthquake Witness the Growth of China's Civil Society [M]. Beijing: Peking University Press, 2009.

[5] Bryan S. Turner. Citizenship and Social Theory [M]. Changchun: Jilin Publishing Group, 2007.

(Translated by Zhang Qian)

Analysis of the Impact of Household Endowment on Responding Capacity of Peasant Households to Disaster Risk[*]

—Based on an Investigation of 39 Poor Ethnic Counties in Southwestern China

Zhuang Tianhui　Zhang Haixia[**]

Abstract: Based on the investigation of 923 households in national poor counties in Sichuan, Guizhou Province and Chongqing City, this paper employed Probit model to analyze the impact of household endowment on household's ability to cope with disaster risks. The research finds that the ability of peasant households against disaster risk was significantly affected by the farmer's family members' experience, the essential time it takes for them to go to the nearest city, the farmer's household area, the income per capita at village level in the surrounding environment, the resource situation and economic organization and so on. Based on those findings, this paper points out

---

[*] This article is funded by National Social Science Fund Western Project "Poverty-stricken Counties of Southwest Ethnic Minorities and Anti-poverty Survey and Assessment" (item number: 09XM008) and the key project of Sichuan Provincial Office of Education "Minority Areas in Sichuan Province Vulnerability of Poor Farmers" (Project ID: 09SA029) and Sichuan Agricultural University "211 Project Phase Ⅲ project". Special thanks and appreciation.

[**] Zhuang Tianhui, Professor of Economics and Management School, Sichuan Agricultural University; Zhang Haixia, teacher of Economics and Management School, Sichuan Agricultural University.

the combination of disaster risk management and poverty reduction is the effective way of enhancing the ability of peasant households to disaster risk. Besides, the paper proposes some potentially important policy implications: enhancing the input of non-agricultural employment training for farmers in minority areas, increasing the technical support and finance insurance to agricultural institutions in these regions; promoting county-level economy in poverty areas.

Key words: Household Endowments; Ethnic Regions; Farmers; Disaster Risk; Responding Capacity

1. Introduction

Natural disasters occurred more frequently and unexpectedly in the 21st century, such as Wenchuan earthquake on May 12, 2008, severe drought in Southwestern China since July 2009, Yushu earthquake in Qinghai Province on April 14, 2010. The main stricken regions are all ethnic minority areas, and rural areas are severely affected ones. Natural disasters seriously hampered the economic development and social stability of minority areas. In addition, crisis awareness, abilities of responding to disasters and tolerance of natural disasters for people in China's minority areas are relatively weak at present, and risk management are mainly conducted by the government, so farmers as the principal role to deal with risks have not been given enough attentions to (Wang Shenghua, Ma Li, 2008). The current system of disaster prevention and disaster coping is mainly concentrated on the building of government's disaster management, while farmers, the main victims, have not been given sufficient attentions to their disaster coping capacity. Under the circumstances of increasingly frequent disasters, farmers begin to play a main role in responding to natural disaster risks. The impact of family endowment on household's ability to deal with disaster risks and its degree or determination factors also become very important. Discussion on these issues has strong realistic significance in improving the minority areas' disaster management system and strengthening the farmers' disaster coping capacity.

2. Literature Review

The study on hazards and disasters has increasingly paid attention to such contributions of households and communities using various resources to cope with natural disasters. (Joost M. E. Pennings and Grossman, 2008, Teun Terpstra and Jan M. Gutteling, 2008, Anwen Jones, Marja Elsinga, Deborah Quilgars, and Jannek Toussaint, 2007). Anderson and Woodrow have emphasized that households' disaster preparedness may be crucial for risk management at the end of the 1980s and the need to identify the capacities that already exist in societies when designing disaster-related development interventions. Mahmud Yesuf and Randall A. Bluffstone (2009) argued that the external circumstances including incomes, the application of loan, the development of labor markets, were very important to households' disaster coping capacity. Anderson and Woodrow (1998) believed that material resources, social organizational structures, motivation and attitude are the main elements affecting the ability to deal with risks. Morrow (1999) sees economic and material resources including personal resources (such as education), family and social resources (such as networks of reciprocity), political resources (such as power and autonomy) and so on. Roger Few (2003) insists that the advanced skills help people respond adversity and take advantage of opportunities successfully. Michael K. Lindell and Seong Nam Hwang (2008) confirms the importance of hazard experience, gender, and income, risk information, and ethnicity in affecting the sense personal risk. Takashi Kurosaki (2006) shows that the ability to cope with negative income shocks is lower for households that are aged, landless and do not receive remittances regularly.

Many scholars in China have studied behaviors of farmers in terms of coping with risks and their strategies. Xu Feng (2000) has discussed the methods of farmers' financial risk management. Ding Shijun and Chen Chuanbo (2001, 2003, 2005, 2007) have concluded that the farmers' risk management strategy can be divided into pre-risk prevention strategies and post-risk prevention strategies, and also have studied farmers' post-risk management measures and their effects as well as family and friends lending non-formal risk-sharing mechanisms. Thus, they have proposed the framework for risk and vulnerability analysis, and integrated risk and vulnerability of rural households into a system of discussion about the risk of various assets, income risks and benefit risks that farmers may suffer from. Qiang Guixia and Nakamoto Kazuo (2008) have established a model of an agricultural business plan by the linear programming, and investigated the agricultural business plan for farmers' diversification and mitigation of risks.

  With further studies on farmers' risk management, the natural disaster risk management has been considered. Wan Jinhong, etc. (2008) has studied drought resilience of rural households by the analysis method of agricultural income diversity. Shuai Hong etc. (2009) has analyzed and evaluated the spatial distribution of farmer's vulnerability in the flood disasters in Dongting Lake area based on study of the flood risk; Cheng Ping, etc. (2008) has analyzed the farmer's risk management principles, strategies, and also the behavior responding mode of agricultural insurance and other risk management tools provided by the government. Zhou Hongjian, etc. (2009) focuses on the relationship among some factors, such as land use, economics of peasant household, cultural qualities and agricultural drought disaster resilience by mathematical and statistical methods. Xie Yonggang, etc. (2007) used disaster economic theories to analyze the impact of natural disasters on farmers' economic situation and their disaster coping capacity via an empirical analysis from the aspect of farmers' income, property damages, injuries, post-disaster liabilities and other aspects. Chen Yuping, etc. (2006) has analyzed the southern drought and its impact on rice production using a set of interview documents of farmers and government departments of southern provinces, and also investigated the treatment strategy of rice drought taken by farmers and the government.

Few documents (researches) specialize in natural disasters in ethnic minority areas. Rong Ning (2007) has revealed that drought and snow disaster have a great impact on western minority areas through the study on natural disasters in western minority areas during 40 years since the founding of People's Republic of China; Wang Shenghua and Ma Li (2008) have discussed the construction of prevention and salvation of agricultural natural disasters in ethnic areas, and also proposed that the government should take the initiative to research and operate agricultural risk management market by market-oriented means to finance relief funds and resources of natural disaster. Xia Jianxin (2008) has proposed an assessment index system of natural disasters' influence on the social-economic and ecological systems, and also given a sustainable development evaluation mode of minority areas under the stress of natural disasters. Using risk management theories, Ha Yudi (2009) has analyzed lessons learned from natural disaster response of current national autonomous areas and the exposed problems during the responding process from the perspective of the government in his essay, namely *Local Government Capacity to Respond to Natural Disaster Management*.

In conclusion: 1) scholars have little specific study on the peasant households' disaster coping capacity, and researches on disaster risk response of farmers mostly focused on risk coping behaviors and coping strategies; 2) studies on risk response of farmers in ethnic minority areas are quite few, while the existing studies on disaster risk management in the minority areas mostly consider disaster risk management from the macro-perspective of the government or the market, which lack enough concerns on farmers who is one of the main roles in response to the risk. This paper investigates the impact of family endowments on the responding capacity of farmers to disaster risk with micro-analysis, which aims to find out the most significant factors in family endowments that affect peasant households' disaster responding capacity, the impacts of different endowments on farmers' ability to cope with risks, hoping to provide a reference for farmers in minority areas to improve the responding capacity of farmers to disaster risks.

3. Hypothesis and Theoretical Model

3.1 Hypothesis

Enhancing the farmers' responding ability to disaster risks is an effective way for disaster prevention and reduction. With the research review at home and abroad, the elements affecting famers' ability to deal with risks include material foundation (such as income, land), family and social networks, family members' awareness of natural disasters, willingness of participation, experiences of disasters, nationality, structure of family, salvation and so on. Rural farmers are the most basic unit of production and consumption. Therefore, their endowments of natural resources and family characteristics constitute the potential risk management capacity of farmers (Chen Chuanbo, 2004). The responding ability of households to disaster risks are mainly affected by the family endowment on the microscopic point of view. Family endowment means that the natural and acquired resources and capacity owned by the farmers' family members and the whole family (Kong Xiangzhi, 2004), specifically including members' health status, education background, personal experience, social networks, resources availability and size of the family business, geographic location and economic environment. Combined with the author's actual survey, the hypothesis is put forward as follows:

Hypothesis 1: The type of household owner, the family members of household, the health status of members, education background, experience and other characteristics of members have impact on the responding capacity of farmers to disaster risks.

Cook (1998) and others argue that the economic and social security of rural households in low-income countries depends greatly on the quantity and quality of household labor and its payment. The endowment of family members is the most important resource and material basis for farmers' responding ability to disaster risks. The better health status of family members', the more abilities to self-saving and reconstruction to reduce disaster losses in the face of natural hazards; members with higher education have more knowledge of disaster prevention and potentials to risk transferring in the case of disasters; well-informed members with rich experience can find more ways to seek for assistance when encountering with disasters, whose responding ability to risks may be relatively stronger. Different nationalities with certain tradition of historical culture affect the cognition, responding ability of households and family members. Here, seven indicators are used for reflecting the peasant household members' endowments, namely "nationality of family members, family structures, the proportion of healthy family members, illiterate family population, the number of migrant workers in the family, if any family members work as village cadres, if anyone in the family has received skills training".

Hypothesis 2: The scales of famers' operation, geographical location and economic conditions have impacts on farmers' ability to respond to disaster risks.

The economic capacity of farmers is the most important capability which directly influences the farmers' ability to cope with risks. Ding Shijun, Sarah Cook (2000) think that the farmland is still an important means to protect the family income and family security; Chen Chuanbo (2003) has shown in empirical studies on farmers' strategies to respond to risks, that the usage of savings, loan, and reducing expenses is the first choice for farmers to cope with large expenditure and economic difficulties, and going out to work is also an important way for farmers to avoid risks. The larger the scale of farmers' business is, the stronger the economic strength may be, and the stronger the ability to cope with disaster risks should be. However, they may also be faced with greater risks because of the large scale, and suffer from greater losses when disasters happen; the more convenient the geographic location of farmers is, the easier the access to a variety of market information and technological information is. Relief may be more readily available when disasters happen, and the responding capability of farmers to risks may be stronger; the better the economic conditions for farmers is, the more savings they may use to overcome temporary difficulties when encountering with risks, and they can get loans more easily due to better loan repayment expectations, which would exert a significantly positive effect on farmers' disaster responding capacity. A total of four indicators in this article are specifically reflected, namely "family of cultivated farmland, the shortest time to reach the nearest town, the family housing area and net income per family member that year".

Hypothesis 3: The technological environment, economic environment and social environment surrounding households have impacts on the responding capacity of farmers to disaster risks.

Environmental endowment of farmers is an important natural resource, and also an important foundation to acquire other resources and capabilities. To most of the native famers, the surrounding environment is unselectable, but it is also one of the important endowments that they own. The better the technological environment for farmers is, the easier they apply technology in reducing disaster losses in face of disaster risks; the better the economic environment for farmers is, the relatively easier the employment and an increase of income would be; the better social environment for farmers, the easier the help from the neighborhood when suffering from natural disasters is. In this paper, it was reflected by a total of six indicators: "the number of family members to receive skills training, whether the village has hold a technical training that year, village-level per capita income levels, whether there is one item or several items of resources in the village (hydroelectricity, mining, tourism), or whether there are professional economic cooperational organizations in the village, the satisfaction with the village cadres, the satisfaction with relations between the neighborhood".

3.2 Theoretical model

The coping ability of farmers to disaster risks is selected as the dependent variable in this article, and it is divided into five levels by the Likert Scale: very weak, weak, normal, strong, very strong. As the dependent variable is the ordinal variable and independent variable is mainly discrete data, it is an ideal estimation method to use the probability model. Ordered Probit model with multiple classes of discrete data is a widely-used method in recent years. Mathematical expressions of ordered probit probability model are referenced to William (2001).

As the actual observed y is discrete, it can not be directly estimated with a linear model. Assuming that the existence of a theory of continuous indicator y^p depends on explanatory variable x. y^p as the unobservable variable, is the mapping of y, which is in compliance with conditions of the ordinary least squares. Therefore, we can mark as follows:

$$y^p = \beta' x + \varepsilon_i, i=1,2,\cdots,n$$

In this equation, β' represents the parameter vector, $\varepsilon_i \sim N(0, \sigma^2 I)$, namely, the observed sample are independent and have normal errors. The existence of boundary points μ_1、μ_2、μ_3 is respectively further assumed as the unknown size segmentation points of households' responding ability to the risk of natural disasters, and there is $1 < \mu_1 < \mu_2 < \mu_3$ as follows:

$$y_i = \begin{cases} 5: \text{if}, y_i^* > \mu_3 & \text{Very strong responding capacity} \\ 4: \text{if}, \mu_2 < y_i^* \leqslant \mu_3 & \text{Strong responding capacity} \\ 3: \text{if}, \mu_1 < y_i^* \leqslant \mu_2 & \text{Normal responding capacity} \\ 2: \text{if}, 1 < y_i^* \leqslant \mu_1 & \text{Weak responding capacity} \\ 1: \text{if}, y_i^* \leqslant 1 & \text{Very weak responding capacity} \end{cases}$$

Probability of $y=1,2,\cdots,5$ are:

Prob. $(y=1|x) = \Phi(-\beta' x)$

Prob. $(y=2|x) = \Phi(\mu_1 - \beta' x) - \Phi(-\beta' x)$

Prob. $(y=3|x) = \Phi(\mu_2 - \beta' x) - \Phi(\mu_1 - \beta' x)$

Prob. $(y=4|x) = \Phi(\mu_3 - \beta' x) - \Phi(\mu_2 - \beta' x)$

Prob. $(y=5|x) = 1 - \Phi(\mu_3 - \beta' x)$

Φ is the cumulative density function of standard normal distribution. Similar to general Probit model, ordered Probit model parameters are estimated with maximum likelihood method. However, the independent variable x on the marginal effect of the probability does not mean coefficient β. As for this probability, the marginal effects of changes in the independent variables are:

$$\frac{\partial \text{prob.}(y=1)}{\partial x} = -\varphi(-\beta' x)\beta'$$

$$\frac{\partial \text{prob.}(y=2)}{\partial x} = [\varphi(-\beta' x) - \varphi(\mu_1 - \beta' x)]\beta'$$

...

$$\frac{\partial \text{prob.}(y=5)}{\partial x} = \varphi(\mu_3 - \beta' x)\beta'$$

Thus, the derivative of Prob. ($y=1$) has obviously an opposite sign with the coefficient β, while the derivative of Prob. ($y=5$) has the same sign with the coefficient β, and the relationship between the derivative of Prob. ($y=2$) and β cannot be determined, but depends on the following measurement

$\varphi(-\beta' x)$ and $\varphi(\mu_1 - \beta' x)$; while the same as prob. ($y=3$) and prob. ($y=4$)①.

This basic model can be set as follows: the responding ability of farmers to natural disaster risks = F (Household member characteristics, household characteristics of the family life means, farmers' environment features) + random disturbance.

4. Data Resource and Variable Setting

4.1 Data resource

Data in this study is supported by western project of national social science fund, which is collected from 39 main minority counties in national poverty reduction, including Sichuan, Guizhou and Chongqing by the "probability proportional to size sampling" (PPS) method from December 2009 to March 2010, with counties as the primary sampling units, accounting for 34.8% of the whole national poverty-stricken

---

① William H. Greene: "*Econometric Analysis (Fourth Edition)*". Beijing: Tsinghua University Press, 2001.

counties in southwestern minority areas of China.

The survey includes 923 households in 160 villages of 104 towns in 39 counties of Sichuan, Guizhou and Chongqing. In order to ensure the survey's quality, the questionnaire survey is conducted experimentally and revised or improved. The formal survey is by means of recruiting investigators, and the seminar recruits senior students whose hometown are located in the surveyed regions from Sichuan Agricultural University, Guizhou University and Aba Teachers' College, and then bring the questionnaires back home in the winter holiday after being trained and examined. The villages are selected from each town, and households are chosen from each village randomly and quantitatively for survey. After the survey, the 1115 questionnaires are collected together, and a total of 923 questionnaires are effective.

The survey shows that only 12% of farmers have strong responding ability to natural disaster risks, while 26% of them have general responding ability to natural disaster risks, and another 62% have weak responding ability to natural disaster risks. Therefore, it can be concluded that in poor minority areas, farmers have weak responding ability to natural disaster risks, which needs to be further improved.

4.2 Variable setting

There are many factors affecting the responding ability of farmers to disaster risks, but the effects of family endowment on the responding ability of farmers to disaster risk are mainly analyzed in this article. Based on the previous studies and hypothesis in this article, combined with field investigations, this article sets family endowment variables affecting the responding ability of farmers to disaster risks from family members' endowment, household economic capacity endowments, family environment endowments. The responding capacity of farmers to disaster risks refers to the nature that farmers are vulnerable to the loss and influence of natural disasters caused by a variety of social and economic factors, which reflects the differences in the responding capacity, buffer, resilience and recovery of farmers to agricultural and natural disasters. The changes of disaster bearing capacity are affected by a series of social and economic factors that are linked to farmers living, production and management (Xie Yonggang, Yuan Lili, Sun Yanan,et al, 2007). Therefore, the responding capacity of farmers to disaster risks is mainly evaluated by the extent that natural disasters affect farmers' production and living in this article. According to Likert scale, the extent that natural disasters affect farmers' production and living is classified into five levels: very weak, weak, general, strong and very strong. Thus, the responding capacity of farmers to disaster risks is also divided into five levels including very strong, strong, general, weak and very weak, which means the stronger the extent that natural disasters affect farmers' production and living is, the weaker the responding ability of farmers to disaster risks is. On the contrary, the weaker the extent that natural disasters affect farmers' production and living is, the stronger the responding ability of farmers to disaster risks is. The extent of natural disasters effects on farmers' production and living is figured out by the questionnaire survey.

Table1 Explanation of model variables and statistical description

Variable Name	Variable Definition	Mean	Standard deviation
Dependent variable			
The responding ability of farmers to natural disaster risks (y)	1 = very weak; 2 = weak; 3 = general; 4 = strong; 5 = very strong	2.336	0.965
Independent variables			
1. Endowment of family members			
If the national type of household owner is minority (NTO)	0 = no; 1 = yes	0.560	0.497

continued

Variable Name	Variable Definition	Mean	Standard deviation
Family structure (FS)			
One couple with two children	0 = other; 1 = one couple with two children (comparison group)	0.305	0.461
One couple with one child	0 = other; 1 = one couple with one child	0.121	0.327
One couple with three children or more	0 = other; 1 = one couple with three children or more	0.185	0.389
One couple, children and grandparents living together	0 = other; 1 = One couple, children, and grandparent with three generations living together	0.251	0.434
Without children or single-parent families	0 = other; 1 = no children or single-parent families	0.138	0.345
Healthy family members (HFM)	the observed value (the completely healthy family members/ total family members)		
Illiterate family members (IFM)	the observed value (illiterate family members/ total family members)		
Egress laboring members (EIM)	the observed value	1.810	1.038
If anyone at home served as village cadres (VC)	0 = no; 1 = yes	0.107	0.309
If anyone at home received skills training in that year (ST)	0 = No; 1 = yes	0.228	0.419

continued

Variable Name	Variable Definition	Mean	Standard deviation
2. Family economic ability endowment			
Family cultivated areas (CA)	the observed value (hm2)	3838.859	5595.613
The shortest time to the nearest market (STM)	the observed value (hours)	0.872	0.762
Family housing area (HA)	the actual observed value (m^2)	119.66	71.44
The family income this year (FIY)	the observed value (RMB)	10151.51	11808.22
3. Environmental endowment			
If the village held a skill training in that year (VST)	0 = no; 1 = yes	0.271	0.445
Economic environment			
Per capita income at village level (VCPI)	the observed value (yuan)	1661	1301.2
Whether there is one item or several items of the village (water, minerals, tourism) resources (RE)	0 = no; 1 = yes	0.476	0.499
Whether there is the specialized cooperative economic organizations in the village (COV)	0 = no; 1 = yes	0.113	0.316

continued

Variable Name	Variable Definition	Mean	Standard deviation
Social environment			
Satisfaction on the village cadres (SCVC)	(1) very dissatisfied (2) less satisfied (3) general (4) satisfied (5) very satisfied	3.706	0.995
Satisfaction on the neighborhood (SN)	(1) very dissatisfied (2) less satisfied (3) General (4) satisfied (5) very satisfied	3.002	0.802

5. Model Estimates and Results Interpretation

5.1 Description of sample analysis

The valid questionnaires on 923 surveyed households shows that permanent population of each household is 4.4 persons. Average household labor is 2.53 persons. Healthy population is 2.88 persons per household. The number of migrant workers per household is 1.8 persons and their education background is shown in Table 2. Distribution of family structure is shown in Table 3. In the surveyed households, 3.14% of the housing structure for peasant families is bamboo huts. 37.7% are the wooden houses. 4.23% are the stone-wood structural houses. 44.53% are the brick-wood structural houses, and 10.4% are reinforced mud. Accordingly, in poor minority areas, family members have lower education background, and their houses are mainly made of brick and wood.

Table 2 Education background of resident population for the surveyed farmers

Education background / Distribution	Illiterate	3 years or less	3-6 years	6-9 years	9-12 years	12 years or more
Proportion /%	23.30	14.28	20.75	23.14	9.48	9.05

Table 3 Distribution of family structure for the surveyed farmers

Family structure / Distribution	Numbers of households	Percentage /%
Single person or one couple	54	5.85
One couple with one child	112	12.13
One couple with two children	283	30.66
One couple with three children or more	171	18.53
Single person with children	42	4.55
One couple and their parents	30	3.25
One couple, children and grandparent with three generations living together	231	25.03
Total	923	100

5.2 Model estimates

With stata10.0 statistical software, data are processed by ordered Probit regression to get the following regression coefficients and test results (Table 4). From Table 4, the log likelihood ratio statistics is −1167.99, and LR $chi2(n)$ is 119.702, while the significance level of log likelihood ratio test is $p = 0.000 < 0.05$, which reflects that the model well fitted the general description, and that impact direction of explanatory variables is basically in line with expected hypothesis.

Table 4 Ordered probit regression results of family endowment on the responding ability of farmers to disaster risks

Explanatory variable	Coefficient	Z value	P value
1. Endowment of family members			
If the nationality of household owner is minority (NTO)	−0.004	−0.051	0.959
Family structure (FS)			
One couple with two children	0.016	0.135	0.892

continued

Explanatory variable	Coefficient	Z value	P value
One couple with one child	0.163	1.513	0.130
One couple with three children or more	−0.086	−0.878	0.380
One couple, children and grandparent with three generations living under one roof	0.165	1.423	0.155
Without children or single-parent families	−0.099	−1.089	0.276
Healthy family members (HFM)	0.105	0.778	0.437
Illiterate family members (IFM)	0.045	1.293	0.196
Egress laboring members (EIM)	0.305**	2.524	0.012
If anyone at home served as village cadres (VC)	0.096*	1.072	0.084
If anyone at home received skills training in that year (ST)			
2. Family economic ability endowment			
Family cultivated area (CA)	−0.018***	−3.406	0.001
The shortest time to the nearest market (STM)	−0.213***	−6.104	0
Family housing area (HA)	0.001*	1.656	0.099
The family income this year (FIY)	0.187***	4.468	0
3. Environmental endowment			
If the village held a skill training in that year (VST)	0.013	0.139	0.889
Per capita income at village level (VCPI)	0.085***	1.568	0.003
Whether there is one item or several items of the village (water, minerals, tourism) resources (RE)	0.363***	4.540	0

continued

Explanatory variable	Coefficient	Z value	P value
Whether there are specialized cooperative economic organizations in the village (COV)	−0.273**	−2.008	0.045
Satisfaction with the village cadres (SCVC)	−0.015	−0.396	0.692
Satisfaction on the neighbourhood	0.039	0.882	0.411
Log *likelihood*			−1167.99
*Pseudo R*2			0.049
*LR chi*2(17)			119.702
Pr *ob*>*chi*2			0.000

Note: ***, ** and * means there is a significant difference at 1%, 5% and 10% respectively.

5.3 Results interpretation

According to the theoretical analysis, if the explanatory variable coefficient is positive, the increase of the value for each variable enhances "very strong" probability of the responding ability of farmers to risks, but reduces "very weak" probability of the responding ability of farmers to risks. Whereas, if the coefficient is negative, it will go opposite. Specific analysis on each factor is as follows:

5.3.1 Effects of family members' endowments on the responding ability of farmers to risks

Based on the estimated results of the investigation, variables reflecting households' experiences in family members' endowments have a relatively significant effect on the responding ability of farmers to risks, such as the number of migrant workers at home, whether or not anyone at home serves as administrative cadre and whether or not family members have received skills training in that year. The trend is basically in line with the realities: the larger the number of migrant workers at home is, the lower the dependence on agriculture is, while the lower the impact on farmers' production and living affected by natural disasters risks is, and the stronger the responding ability of farmers to natural disasters risks is. If anyone at home serves as village cadre, then their social resources may be more abundant than other farmers, so they will have more responding methods to risks and relatively strong responding ability to risks. Compared with other households, households with members who have received training have stronger responding ability to disaster risks. Taking southwestern drought from 2009 to 2010 as an example, in Zunyi County of Guizhou Province, villagers succeeded in fighting against drought and keeping seedlings in spring in the local minority areas which suffered from severe drought by the promotion of "shallow dry nursery planting" method to reduce the potential economic losses caused by drought.

Meanwhile, the results of the investigation is inconsistent with the expected assumption, which means there is no significant difference in effects on the responding ability of farmers to natural disasters risks by health condition and education background of family members. This is probably because the natural disasters risks have dynamic variation and complex diversity, and there is no particularly significant difference in effects on the responding ability of farmers to natural disasters risks by health condition and education background of family members. In addition, the effects of family structure on the responding ability of farmers to natural disasters risks are not significant, but its impact direction can be seen from positive or negative factors. Compared with the comparison group (one couple with two children), the responding ability of farmers with "one couple, children and grandparent with three generations living under one roof" family structure to natural disasters risks tended to be lower.

5.3.2 Effects of family financial capacity on the responding ability of farmers to natural disasters risks

According to the results of the investigation, index reflecting the potential economic strength of households in family financial capacity endowments has relatively significant differences in the responding ability of farmers to natural disasters risks, such as family cultivated areas, the shortest time it takes to get to the nearest county and family housing areas. The results show that family arable land areas have an extremely significant impact on the responding ability of farmers to natural disasters risks, and the impact direction is negative, which is in line with the expected hypothesis. In ethnic minority areas, agricultural technology is underdeveloped, so the agricultural production mainly depends on the amount of arable land. The larger the cultivated area is, the relatively more agricultural income is, but in minority areas with frequent disasters and poor agricultural science and technology, the larger the arable land is, the greater the disaster risks for farmers to cope with are. Once suffering from natural disasters, farmers would have poor production or even no production, which is basically consistent with the reality that farmers in minority areas fall into poverty rapidly after natural disasters. The effect of the shortest time it takes to get to the nearest county on the responding ability of farmers to natural disaster risks is negative, which is in line with the expected hypothesis. The longer the shortest time it takes to get to the nearest county is, the longer the distance to city is, which means the transportation is inconvenient and traveling vehicles are underdeveloped. In ethnic minority areas, farmers and herdsmen are scattered around the country with poor traffic conditions, so counties are usually the local economic and cultural center, and also the main center of agricultural products trade. Farmers who are closer to counties have more ways to occupy non-agricultural production. Furthermore, farmers who are closer to counties are easier to get public resources, such as market information, technological services and medical services, thus they all have more advantages than other families. The housing areas of farmers reflect the farmers' economic strength to some extent, since it is the effective performance of their long-term wealth accumulation. In general, farmers with larger housing areas have better economic strength not only in that year but also in recent years. This index objectively reflects that farmers' economic strength has a significant impact on the responding ability of famers to natural disaster risks.

The results of the investigation show that per capita net income of households in that year has significant impact on the responding ability of famers to disaster risks, which is not in line with the expected hypothesis. Investigations suggest that per capita net income of the sampling farmers in deep poverty in that year is less than 1,196, accounting for 76.6 percent. This income level can only meet the food and clothing demands, and it is difficult to resist natural disasters. According to rural poverty monitoring results from the survey by National Committee on National Autonomous Areas, by late 2007, poor population in rural areas of National Autonomous Areas is 7.736 million, accounting for 52.3% of the poor population in rural areas. Low-income population in this area account for 52.1% of low-income population in the whole country with a poverty rate of 6.4%, which is 4.8 percentage higher than that of the national average. To conclude, most farmers in ethnic minority areas are in deep poverty, and there are also a lot of low-income households. When a large number of poor peasant households suffer from natural disasters, their economic ability to cope with natural disaster risks is extremely weak, which may explain the results that the level of income has great impact on the responding ability of farmers to disaster risks.

5.3.3 Effects of environmental endowment on the responding ability of farmers to disaster risks

Seen from the results of the investigation, family and economic endowment have relatively significant impact on the responding ability of farmers to disaster risks. Especially, three economic environment endowments including "per capita income level, resources, and economic organizations at the village level" have a relatively significant impact on the responding ability of farmers to disaster risks. Whether there is one or several advantages such as water, minerals, tourism resources has a positive effect on the responding ability of farmers to disaster risks. From the results, farmers who have one or several advantages such as electricity, mineral, tourism resources have stronger responding ability to disaster risks. This is possibly because farmers who have electricity, mineral, tourism resources can obtain many non-agricultural ways to get employed. In such circumstance non-agricultural income may become their main source of income, then the collective economic strength of their village would be relatively stronger, so natural disasters have less impact on them, and their responding ability to disaster risks is relatively stronger.

Meanwhile, seen from the results, satisfaction of households to village cadres and neighborhoods has no significant impact on the responding ability of farmers to disaster risks. Investigation shows that the average satisfaction of the surveyed farmers to villages cadres and neighborhoods is about "3", which means general. 72.8% of the surveyed farmers in the villages have not received technical skills trainings, which also explains the reason why farmers' technical environment has no extremely significant impact on their responding ability to disaster risks to some extent.

6. Conclusions and Policy Implications

6.1 Conclusions

Based on the actual survey data of households, the effects and differences of family endowments on the responding ability of farmers to disaster risks are analyzed by ordered Probit Probability model in this study. The results show:

Firstly, the experience endowment of family members is an important factor that affects the responding ability of farmers to disaster risks. Whether there are any family members who serve as village cadres and receive skill trainings have a significant effect on the responding ability of farmers to disaster risks. It also shows that other non-agricultural employment ways such as increasing the number of migrant workers are essential for farmers to enhance their responding ability to disaster risks.

Secondly, the family's economic strength is key to affect the responding capacity of farmers to disaster risks. The larger the family's arable land area is, the weaker the responding ability of farmers to disaster risks may be, and the responding ability of households who are closer to counties to disaster risks may be stronger. Famers with larger family housing areas have stronger economic strength, and their responding ability to disaster risks is also stronger.

Thirdly, the economic environment of farmers has a great impact on the responding ability of farmers to disaster risks. Per capita income, resources and village economic organizations at village-level have a significant effect on the responding ability of farmers to disaster risks, and they all have positive effects, which indicates that good economic environment is helpful for farmers to enhance their responding ability to disaster risks.

6.2 Policy implications

We believe that these findings have some potentially important policy implications.

Firstly, enhancing the intensity of anti-poverty is the fundamental way of improving the peasant households' responding ability to disaster risks in ethnic counties. The study result suggests that the economic environment of farmers has a great impact on the responding ability of farmers to disaster risks. In the ethnic areas with frequent natural disasters, it should effectively integrate anti-poverty, disaster risk management and improvement of the capacity of farmers, and enhance the responding capacity of households to disaster risk through increasing farmers' income.

Secondly, the combination of disaster risk management and poverty reduction is the way of enhancing the responding capacity of households to disaster risk. The input of non-agricultural employment training and transferring for farmers in minority areas should be enhanced. The efforts of farmers' skill training should be strengthened, especially agricultural skill training in disaster prevention. Meanwhile, the skill training of egression farmers should be increased, and policy assistance should be provided for potential migrant farmers so as to increase employment channels for migrant workers.

Thirdly, the technical support and finance insurance to agricultural institutions in these regions should be increased. The technological guidance for famers who are engaged in agriculture for a long time, especially the planting peasants, should be strengthened to improve the quality of agricultural products and promote agricultural products to the market. At the same time, the planting peasants should be given financial support to strengthen the construction of agricultural insurance and enhance their responding ability to disaster risks.

Fourthly, ethnic regional resources should be sustainably developed by economic organizations at the village-level, industrial platform and market access should be provided for farmers to increase income continuously. Investment and construction in communications, transportation, health, education and other infrastructure investments in minority areas should be enhanced. In terms of policy and system design, both individuals and enterprises are encouraged to participate in the investment and development of the local electricity, mineral, agricultural products and special tourism resources. Meanwhile, the sustainable development and protection of ethnic regional resources should be guided. Combined with the actual development of the ethnic economy with the local characteristics, construction of the specialized cooperative economic organizations should be actively encouraged, so that the rural collective economic strength in minority areas will be enhanced.

References:

[1] Rong Ning. Preliminary Study of Natural Disasters in the Western Minority Areas for 40 years Since the Founding of China [J]. Qinghai Ethnic Studies, 2007.

[2] Wan Jinhong, Wang Jingai, Liu Zhen, etc. Drought Resilience of Rural Households from the Perspective of Income Diversity—Taking Xinghe County of Inner Mongolia as an Example[J]. Natural Disasters, 2008.

[3] Shuai Hong, Liu Chunping, Wang Huiyan. Assessment on Flood Disaster Vulnerability of Farmers in Dongting Lake Area [J]. Natural Disasters, 2009.

[4] Cheng Chengping, Liu Suchun. Agricultural Risk Management Strategy Based on Farmers Perspective [J]. Contemporary Economic Management, 2008.

[5] Zhou Hongjian, Wang Jing, Jia Huicong, etc. Factors of Agriculture

Drought Disaster Resilience—Based on Field Measurements and Household Survey of Land Use [J]. Yangtze River Resources and Environment, 2009.

[6] Xie Yonggang, Yuan Lili, Sun Yanan. Analysis of Impact of Natural Disasters on the Economic Status of Farmers and Their Bearing Ability to Disasters [J]. Natural Disasters, 2007.

[7] Chen Chuanbo, Ding Shijun. Analysis of Households' Risk and Treatment Strategies [J]. China's Rural Economy, 2003.

[8] Xu Feng. Family Financial Risk Management [J]. Agricultural Techology and Economy, 2000.

[9] Chen Fengbo, Chen Chuanbo, Ding Shijun. The Risk and Management Strategies of Drought for Farmers in Southern China [J]. China's Rural Economy, 2005.

[10] Chen Chuanbo. Empirical Study on the Informal Risk-sharing of Farmers in China [J]. Agricultural Economy, 2007.

[11] Ding Shijun, Chen Chuanbo. Household Risk Management Strategy Analysis [J]. Agricultural Modernization, 2001.

[12] Chen Chuanbo. Farmers' Risk and Vulnerability: An Analytical Framework and the Experience of Poverty-stricken Areas [J]. Agricultural Economy, 2005.

[13] Li Gucheng, Feng Zhongchao, Zhan Shaowen. Impact of Family Endowments on Family Technical Efficiency in Households-Based on Empirical Study on Stochastic Frontier Production Function of Famers in Hubei Province [J]. Statistical Research, 2008.

[14] Kong Xiangzhi, Fang Songhai, Pang Xiaopeng, etc. Effects of Household Endowments in Western China on Agricultural Technology Adoption [J]. Economic Research, 2004.

[15] Anderson, M. B., Woodrow, P. J. Rising from the Ashes: Development Strategies in Times of Disaster. London: Intermediate Technology Publications, 1998.

[16] Joost, M. E. Pennings, Grossman. Responding to Crises and Disasters: The Role of Risk Attitudes and Risk Perceptions[J]. Disasters, 2008, 32(3): 434-448.

[17] Teun Terpstra & Jan M. Gutteling. Households' Perceived Responsibilities in Flood Risk Management in the Netherlands. Water Resources Development, 2008 (4): 555-565.

[18] Anwen Jones, Marja Elsinga, Deborah Quilgars, Janneke Toussaint. Home Owners' Perceptions and Responses to Risk[J]. European Journal of Housing Policy, 2007,7(2):129-150.

[19] Mahmud Yesuf and Randall A. Bluffstone. Poverty, Risk Aversion, and Path Dependence in Low-Income Countries: Experimental Evidence from Ethiopia[J]. Amer. J. Agr. Econ. 2009,91(4): 1022-1037.

[20] Morrow, B. H. Identifying and Mapping Community Vulnerability[J]. Disasters, 1999(1):1-18.

[21] Roger Few. Flooding, Vulnerability and Coping Strategies: Local Responses to a Global Threat South Bank University, London, UK[J]. Progress in Development Studies,2003(1): 43-58

[22] Michael K. Lindell, and Seong Nam Hwang. Households' Perceived Personal Risk and Responses in a Multihazard Environment[J]. Risk Analysis, 2008 (2):539-556

[23] Takashi Kurosaki. Consumption Vulnerability to Risk in Rural Pakistan[J]. Journal of Development Studies, 2006(1):70-89

Strengthen Capacity of Poverty Reduction by Village Public Products Supply

—Taking Jiezhu Village, Xi'eluo Town of Yajiang County in Ganzi Tibetan Autonomous Prefecture of Sichuan Province as an Example*

Li Xueping Longming Azhen

Abstract: Through strengthening the village public products supply to enhance the villagers' capability could achieve poverty reduction. Based on the theoretical context of equal access to basic public services and the concept of capability, and the case studies of Jiezhu village, this paper puts forward that to improve the capability of impoverished population in concentrated special type poverty-stricken areas is the fundamental goal of poverty reduction. To realize this goal must depend on the all-round "breakthrough" of the capability from survival to

---

* Supported by the 2009 Ministry of Education on Philosophical Social Science Research of key research project "Urban and Rural Social Governance Research"(09 JZD0025); supported by the 2008 Ministry of Education of Humanities and Social Science Research base on major project "In the Process of Urban and Rural Social Management System Reform Research" (08 JJD810156); supported by 2010 Huazhong Normal University independent scientific research project "Poverty Alleviation and Development Research in Concentrated Special Type Poverty-stricken Areas (Wuling mountainous area)".

Li Xueping is a professor at Sociological School, Social Development and Policy Research Center of Huazhong Normal University; Longming Azhen is an officer at Ganzi Tibetan Autonomous Prefecture in Sichuan Province.

production and then to development. This paper focuses on improving capability, and pays more attention to supply intangible village public products such as communicational ability and practical technical training, etc. The poverty relief in concentrated special type poverty-stricken areas lays more stress on the responsibility of government, and even on the responsibility of the central government. The idea and practice of poverty reduction should be embedded in the national traditional culture and value concept to realize the endogenous development.

Key words: Village; Public Products Supply; Capability; Poverty Reduction; Concentrated Special Type Poverty-stricken Areas

At present, the research on rural public product supply and poverty reduction in academic circles mainly focused on farmers' income growth. One consensus has been formed that the rural public product supply can alleviate poverty in many ways, including direct alleviation and indirect alleviation. For example, it can increase the farmers' income and reduce expenditure at the same time. In the research from the perspective of income, it seemed to be more subtle to explore the rural public product supply's influence on poverty reduction from the perspective of basic guarantee for farmers[1]. The research mainly focused on the western China and some other provinces. For example ZhaoXi[2] and Fan Shenggen discussed the poverty reduction model in western areas; Liu Liu[3] and Peng Xinglian[4] discussed the function of rural public

---

[1] Xu Yi. *The Imperfection and Reform Ideas of Rural Public Product Supply System*. Journal of Anhui Technical Normal College, 2005, Vol, 2.

[2] Zhao Xi, Yan Hong & Xiu Huiling. *The Model Study of Development-Oriented Poverty Reduction Strategy in Western Rurual Areas*. Inquiry into Economic Issues, 2007, Vol, 12.

[3] Liu Liu. *The Rural Public Products Supply in Guizhou to Ease Poverty Based on the Study on the Impact of Poverty Relief Funds*. Master's Degree thesis of Guizhou University, 2008.

[4] Peng Xinglian. *Rural Public Product Supply and Jiangxi Farmers' Income*. Master's Degree Thesis of Nanchang University, 2007.

production supply for poverty reduction in Tibet, Guizhou and Jiangxi. But the research on case study of a specific village was quite rare. As for the researches' content, most of them discussed the status of rural public production supply, supply structure, system construction, and the existing problems and countermeasures, such as the research conducted by Sui Dangcheng[1] and Shao Guiwen[2]. Generally speaking, the feature of research on the relationship between rural public products supply and poverty reduction is macro-research object and huge analytical framework so far. This paper takes Jiezhu village, Xi'eluo town of Sichuan province as an example, from the perspective of the famers' capability to explore the role that rural public product supply plays in improving the capability to achieve poverty reduction. On July 7 to 17, 2009, the author carried out investigation in Xi'eluo County and Jiezhu village and gathered all the materials from this investigation. The specific data concerning Xi'eluo County and Jiezhu village provided by Xi'eluo government and Poverty Relief Office of Yajiang County.

1. Theoretical Context

In the late 18th century, scholars defined poverty mainly according to "food consumption". For example, Rowntree distinguished poverty from non-poverty on the basis of whether "physiological efficiency" of income has met or not. Until 1965, Qusanski determined minimum food spending and applied specific engel's index to differentiate poverty from non-poverty, and put forward poverty depends on the level of income, that is to say, whether possess a certain amount of money to meet the basic needs of human[3]. Qusanski's methods are still widely used by many scholars, nations and international organizations. Amartya Sen

---

[1] Sui Dangcheng. *Rural Public Product Supply Structure*. Doctoral Dissertation of Northwest A & F University in 2007.

[2] Shao Guiwen. *Rural Public Goods Supply Institutional Research*. Master's Degree thesis of Central South University in 2007.

[3] Feng Ying. *The Evolution of Poverty Definition and Thinking about Poverty Issues in China*. Economic Research Guide, 2010, Vol,18, P. 6.

expanded the concept of poverty from income poverty to right poverty, capability poverty and human poverty, and enlarged the cause analysis of poverty from economic factors to politics, law, culture, system, etc. He elaborated on capability poverty and governance in *Development as Freedom*. Sen illustrated the concept of capability from the two angles: microscopic individuality and macroscopic totality. In Sen's opinion, substantive freedom was a kind of capability.

Starting from the microscopic individuality, Amartya Sen defined poverty as the capality to achieve functional activity. In his view, "when analyzing social justice, there are strong reasons for us to judge the situation of someone by the capability the person has, that is the substantive freedom that the person possesses and the freedom to enjoy the life he cherishes. According to this perspective, poverty must be considered as deprival of basic capability, rather than low income. One's capability refers to the functioning combination which possibly achieves for the person. Therefore, capability is a kind of freedom, which is the substantive freedom for achieving kinds of possible functional combination". "The concept of 'functioning', reflects the various things or status which a person considers to be worth doing or achieving. There are many kinds of valuable functionings from the very primary request such as obtaining enough nutrition and exempting from the disease that can be avoided to quite comprehensive functionings or individual status, such as participating in community life and obtaining self-esteem." "A person's actual achievement can be presented by a functioning activity vector. A person's 'capability set' consisted of the mutual replaceable functioning activity vectors which the person may choose. Therefore, a person's functioning combination reflects the actual achievements he has reached, while the capability set represents the freedom which the person may freely realize."[1] During the specific study, Sen evaluated equality from the perspective of 'the life content

---

[1] Amartya K. Sen. *Development as Freedom*. Renmin University of China Press, 2002, Vol, 85, PP. 62-63.

domain' and the ability to achieve the life content, and further elaborated the measuring methods of ability and the multiple subjects' role in the ability to implement equality[①], constructed the concept of equality from ability perspective. Inspired by the poverty theory and method from Sen, the United Nations Development Programme respectively designed "capability poverty index" and "human poverty index" in 1996 and in 1997. The basic content of the two poverty indexes were both related to the real life and freedom that people had possessed. [②]

From the macroscopic totality perspective, Sen regarded freedom as value orientation, and regarded increasing personal benefits as value goal. He considered that freedom was not only the primary purpose of development, but also the indispensable important means to promote development. The freedom which Sen referred to means the capability of enjoying the life that people have reason to cherish. In practical terms, "substantive freedom including basic capabilities of escaping from hardship such as: hunger, malnutrition, avoidable diseases, premature death, and the freedom of literacy, enjoying political participation and so on"[③]. Freedom plays a constructive role in development: freedom is the inherent part of people's value standards and developmental goals, it

---

[①] Liu Deji. Amartya K. *Sen's Views on Equality of Capabilities and Equality of Public Services*. Shanghai Economy Research, 2009, Vol 11, PP. 110-109.

[②] "Ability poverty index" is a composite index, which is consisted of three indicators. The three indexes refer to: the proportion of children underweight below 5 years old; the proportion of children born without professional health care personnel, the proportion of illiterate women under the age of 15. The three indicators get a mean according to equal weighting the aggregation that is the ability poverty index. "Human poverty index" is composed by three index component: life deprivation, knowledge deprivation and living standard deprivation. Life deprivation index in developing countries points to the death proportion of the population before 40 years old, while in developed countries it points to the death proportion of the population before 60 years old; knowledge deprivation index is the illiteracy rate among adults; living standards deprivation is a comprehensive index, which includes three aspects: the proportion of population who are unable to get access to medical services, the proportion of population who are unable to get safe drinking water, the proportion of malnourished population under the age of 5.

[③] Amartya K. Sen. *Development as Freedom*. Renmin University of China Press, 2002, P. 30.

is valued by itself, and therefore it does not need to present its value by the connection with other valuable things, neither standing out by playing a prompt role in other valuable things. At the same time, freedom is still the principal means of development. The most important five instrumental freedoms include political freedom, economic condition, social opportunity, transparency guarantee and protective security, which can help people live more freely and improve their overall capability in this aspect.

Sen constructed concept of capability, and also constructed the concept of ability poverty governance. That is to say, ability poverty presents as low capability. And the enhancement of capability relys on social arrangement, but the social arrangement at least needs to perfect the five instrumental freedom. Academic circles hold that: to a certain extent, the development concept of Sen is the legitimate argument for the equalization of public service in social construction in our country. Freedom and basic public service both have the constructive significance and instrumental significance. The constructive significance of equalization of public service exists in public products supply meeting the social public need, increasingly achieving the social and public interests, and ensuing the existence and sustainable development of human society to demonstrate the public value. There is inherent connection that happens to coincide with the instrumental significance of equalization of public service and the five instrumental freedoms Sen stressed[1]. At least the construction of five instrumental freedom of Sen contains supplying various kinds of public production to people, especially those in backward areas and vulnerable groups. In the actual social practice in our country, as for supplying public products to the backward area and impoverished population, the fundamental idea is to achieve the equalization of public service. In order to achieve

---

[1] Zeng Jingjing. *A New Interpretation of Equalization of Basic Public Services: A Perspective of the Free Development of Amartya Sen.* Preceedings of Nanjing University of Technology (Social Science Edition), 2010, Vol.2, P.54.

equalization of public service, one of the essential public policies is "filling the short board", namely the public products supply should show preference to the backward area and vulnerable groups.

Each kind of theory explaining the occurrence of poverty provides one or more factors and ways that influence poverty alleviation. However, public products supply is involved in all of the theories, for people believe that public products supply can perfect the poor welfare. ① The first report comprehensively analyzing the mutual relationships between the rural poverty status and public product supply in China was accomplished by the World Bank in 1992. The study holds that public products supply in the poor areas is of positive and significant efficiency and fair meaning. ② We consider that: If China's current economic social development has entered into the public products supply era, then poverty reduction will have especially entered into the time of strong public products supply. As "National Program for Rural Poverty Alleviation (2001—2010)" points out: "The central and local governments financial relief fund, must be mainly used in perfecting the basic production and living conditions and the construction of infrastructures."

Through the public products supply to solve the problems caused by social interest groups differentiation and interest differentiation of areas, is a general approach among the world, and China is no exception. The reason is that: First, products supply and distribution play an important role in reducing income disparity. It should be said that, it plays a bigger and more direct role in supplying balanced public productions to the backward areas and vulnerable groups. As Anand and Ravallion considered that, on the spending structure the government should take the investment in public services as priority; in

---

① Liu Liu. *Rural Public Product Supply and Alleviate Poverty*. Journal of Guizhou Provincial Committee Party's School of CPC, 2009, Vol, 6, P. 5.

② Liu Liu. *The Impact of Rural Public Products Supply in Guizhou to Alleviate Poverty Based on the Structure Analysis for the Poverty Relief Funds*. Master's Degree thesis of Guizhou University in 2008, P. 4.

spending fields, government should look after low-income groups and poverty areas at first, because these fields are the places where maximum marginal utility lies in. ① Public products supply can directly improve the life conditions in backward areas and vulnerable groups in many ways, enhance the capability of people, and realize the substantive freedom of people. "Better education and health care not only can directly improve the quality of life, but also can raise the capability to obtain income and get rid of poverty at the same time. If education and health care become more popular, if would be more likely for those who were poor to get better opportunity to overcome poverty."② The healthy body brought by public health and medical service can enable people to live a high-quality life; education plays an important role in cultivating skills, making people aware of and grasp the opportunities, protecting impoverished population from the vicious circle: low income—low education investment—low capability—low income; infrastructure improvements will directly improve living standards and the quality at the same time, expand people's capability in aspects of production and exchange. That is to say, public products supply assumes the role of "protection" and "promotion". The former focuses on preventing people from being reduced into lower living standards, and the latter aims to improve living standards and expand capability.

In short, the relationship between public products supply, capability and poverty reduction can be summarized as following:

$$\text{the public products supply} \xrightarrow{\text{aim to}} \text{raise capability} \xrightarrow{\text{to realize}} \text{poverty reduction}$$

2. Public Products Supply in Jiezhu Village

The article took Jiezhu village as an example, which lies in Xi'eluo

---

① China (Hainan) Reform Development Research Institute Group. *The Realization of Human's Comprehensive Development——The Basic Public Service and China's Human Development*. China's Economic Press, 2008, P. 24.

② Amartya K. Sen. *Development as Freedom*. Renmin University of China Press, 2002, P. 88.

town, Yajing County (key counties for national poverty alleviation and developmental work), Ganzi Tibetan Autonomous Prefecture of Sichuan Province (concentrated special type poverty-stricken area), where 98 percent of population are Tibetan. By the end of 2009, the per capita net income of farmers in Xi'eluo town was 1,879 yuan. The poverty population totaled 1,223 and the poverty rate was 38%. To conclude, Xi'eluo town and Jiezhu village with deep poverty and high poverty rate display the typical characteristics of concentrated special type poverty-stricken areas.

2.1 The production and living subsidies for Jiezhu village residents

When Sen investigated the poverty, he expanded low income to low capability, but he did not negate relationship between low income and poverty, he proposed: "low income can be the important reason for deprivation of a person's capability. Income shortfall really is a strong inducing condition that cause poverty." "The relative deprivation of income can produce absolute deprivation of capability." And "between the income deprivation and the difficulty in changing income to functionings, there is a certain coupling effect. The defects in capability such as age, disability, illness, will lower the capability to obtain income. And these factors also make transforming income to capability more difficult. As for the person with older age, more serious disability, and more severe illness, they will need more income(in order to get care, correct disability, receive treatment) to accomplish the functionings compared with others. This determined that the 'real poverty' according to capability deprivation, may be more severe than the poverty showed by income opportunities at some obvious level"[1].

Considering the relationship between the low income and capability in the village of concentrated special type poverty-stricken areas such as Jiezhu village, improving the villagers capability means that increasing the income of the villagers is still the primary task. Prividing all kinds

---

[1] Amartya K. Sen. *Development as Freedom*. Renmin University of China Press, 2002, PP. 86-85.

of subsidies is the most direct approach for income growth. In recent years, the government has provided various kinds of subsidies as followings.

2.1.1 Medicaid

The whole villagers have participated in the New Rural Cooperative Medical System. The fund for cooperative medical is 100 yuan each year. Each individual hands in 20 yuan, and local governments pay another 80 yuan for each villager. Since Yajiang is a national-level poverty-stricken county, all of the 80 yuan is paid by the central finance.

2.1.2 The minimum living allowance

There are 191 people who enjoy subsistence allowance in Jiezhu village. The standard is 45 yuan per month per person.

2.1.3 Food direct subsidy and comprehensive direct subsidy

These subsidies should be distributed to the grain famers. Food subsidies mainly include the seed subsidy, and the subsidy standard is 10.77 yuan per mu; in comprehensive direct subsidy, the subsidy is 100.95 per mu. So the total subsidy amounts to 111.72 yuan per mu. The larger the area for planting grains, the more subsidies.

2.1.4 Returning the grain plots to forestry subsidy

The subsidies converted into RMB is on average 260 yuan per mu each year, and the subsidy has last for 8 years. The specific distributing method of subsidies according to the need of villagers, is to issue 130 yuan per mu, and issue the grains which value 130 yuan per mu. Returning the Grain Plots to Forestry would be checked and accepted in 2009, if the acceptance was passed, villagres would continue to accept aid for another eight years; if acceptance was unqualified, villagers should return all the subsidies and get fined. Xi'eluo county has passed the acceptance, and the government will continue to aid farmers for 8 years.

2.1.5 Allowance

The village picks out for the poorest families each year, and the subsidy is 195 yuan per person. The population who get the subsidy is not stable. There are more people who get the subsidy and more subsidies from the county during the disaster year. The total amount is above 50,000 yuan in general, sometimes up to 70,000 to 80,000 yuan.

2.1.6 Family planning aid

The kind of aid include: (1) "fewer children equals faster prosperity" subsidy is aimed at the married women under the age of 49, who only have two children and take birth control measures, and one-time subsidy is 3,000 yuan. There were about 20 families which get this subsidy in Jiezhu village in recent years; (2) mutual assistance awards are aimed at the old above the age of 60, who only have one children or only two daughters; after passing the application, everybody can get subsidies of 600 yuan per year, and one couple get 1200 yuan per year. The subsidy object can get the money until death. (3) one-child allowance. The couples, who volunteer to have just one child, can get 10 yuan per month until the child grow up to the age of 18.

2.1.7 "The three old cadre" subsidies

"The three old cadre" refers to village cadres, activists and party members who are in position during the years from 1959 to 1962. The allowance standard is 180 yuan per month per person, and in addition there are another 500 yuan every year for Spring Festival. "The three old cadre" can get this subsidy until death.

2.1.8 Civil administration subsidies

(1) Famine relief. In disaster year, there are 2,000 yuan relief funds for spring famine in the village, mainly used to buy seeds. (2) The medical relief. This relief mainly involves low income households. When they see the doctor besides medicare reimbursement, through application, approved by township government, they can get reimbursement of 100 yuan to 5,000 yuan at County Civil Affairs Bureau. (3) Disaster relief. There will be these subsidies when large disasters happen, but the amount of grants is not stable.

In terms of the actual effect of subsidies delivery, it has directly raised the income of villagers, and strengthened their capability. The delivery of living aids undoubtedly generated effect, and the effect is also obvious in production subsidies issued. Academic circles considered that: compared with "secret subsidy" which aid the grain production enterprises, the "apparent subsidy" which is issued to grain farmers in cash, from the perspective of integration production, the latter provides early capital for production and results in better effect in enhancing the villagers' enthusiasm for production and income. The reason is that farmers should produce grain before obtaining "apparent subsidy". Therefore the effect of "secret subsidy" fades compared with "apparent subsidy"①.

In addition to all sorts of direct subsidies, another approach to increase the cash income of villagers is the implementation of work relief in infrastructures. In recent years, the implementation of work relief project and the expenditure on Xi'eluo county include: relocating 100 households; investing about one million yuan in work relief, more than 200,000 yuan in work relief to build Zha'a bridge and about 300,000 yuan in work relief to construct concrete road in Jiezhu village.

2.2 Infrastructure construction

Village infrastructure construction includes road construction, human and livestock drinking water facilities construction, power supply, radio and TV facilities construction, communication facilities construction, etc. The study from Sui Dangcheng showed that in terms of contribution rate, the infrastructure construction investment in rural areas made the biggest contribution rate for per capita net income of rural residents②. Moreover, infrastructure construction can directly improve the living conditions of villagers, and enhance the capability of villagers. For example, the benefits of road construction in favour of strengthening the capability of villagers are as follows: First, improve the villagers travelling conditions, make job hunting easy, and save transportation cost, etc. Second, expand the scope of communication,

---

① Jin Shuanghua. *The Public Products Supply and the Income Gap between Urban and Rural Area*. Journal of Dongbei University of Finance and Economics, 2008, Vol,5, P. 47.

② Sui Dangcheng. *Rural Public Product Supply Structure*. Doctoral Dissertation of Northwest A & F University in 2007, PP. 106-105.

increase village social capital. Third, develop the market, enable agricultural products to be transported in time, and then realize its commodity value; perfect agricultural trading conditions and reduce the transportation cost. As another example, village broadcasting and TV, communication network construction are conducive to reducing productional cost, transactional costs and market risk. Economic activities include the production activities as well as trading activities. The cost generated in the former is called productional cost, and the latter named transactional cost. The reason for transactional costs is the information asymmetry. Favorable village public products supply can enable the farmers to have more access to trading information, so that finding transaction objects, bargaining, and winning trading contract will be easier.

Under the comprehensive development-oriented poverty reduction model in village, the situations of infrastructure construction in Jiezhu village are like that: the project of expanding radio and television coverage in the rural areas has been accomplished, and television is widely available to each household; water pipe were equipped in every household in 2008; a China mobile base station and China unicom base station have both been constructed in Jiezhu village, and mobile signal coveres all over the village; there are abundant water and electricity in whole Yajiang county, and with adequate supplies, every household has access to electric light. In recent years, the construction of roads and bridges in Xi'eluo town has enabled every household to have access to highway (including service roads towards villages). These constructions include: (1)During the year 2007 to 2009, the government invested 20 million yuan to construct the Niu Xi Road of about 13 kilometers from 318 National Highway to Yajiang county. (2) In 2008, the government invested 800,000 yuan to build cement road covering 1 km in Jiezhu village. (3) In 2008, the government spent 600,000 yuan constructing a bridge in Jiezhu village.

2.3 Basic public services

The supply of health care and basic education can greatly improve the capability of impoverished population, including perfecting daily living conditions and enhancing the population's quality. Xi'eluo town is equipped with one hospital, two health clinic rooms, one primary school at county level, three primary school at village level. There are altogether 399 students and 20 teachers. Jiezhu village is the seat of Xi'eluo town goverment. Primary school, clinics and hospital directly serve residents in Jiezhu village.

The public medical care service which Jiezhu village residents enjoy include: First, all of Jiezhu village residents have participated in the New Rural Cooperative Medical Care System, and part of the villagers even can get medical relief. Second, township hospital is located in Jiezhu village, so the villagers can receive medical service in village. Third, in terms of the village environmental construction, township government has raised funds and built a rubbish collection station in Jiezhu village, and the township government hired a cleaner to sweep the street. When the author surveyed in Jiezhu village, he observed that there were no other pollutants except cow muck, especially no white garbage flying in the streets or around the village.

In terms of education, the government has implemented a policy for basic education that government pays the accommodation, meals and tuition, the entire school-age children in village can accept nine-year compulsory education for free. The enrollment rate is 100% and dropout rate is zero.

2.4 Living services

In terms of living services, in order to improve the villagers' living conditions, the public products supply which the government at all levels provide mainly includes: the housing construction (rebuilding), supply for alternative energy facilities, protection for village community safety.

(1) Housing construction. (a) In recent years, the Xi'eluo town government has already invested 730,000 yuan to help 83 households alleviate the housing difficulties. (b) 60 households have accomplished housing reconstruction, and the investment is 1.8 million yuan.

(2) Alternative energy infrastructure. The government has designed and produced energy-saving stoves, and distributed them to villagers for free; besides, the government provided 20 set of energy lighting equipments with 20 watts.

(3) Village security. Ensuring village safety shows the interactive cooperation between government and villagers. The specific measures are implemented by joint security defense, the main contents include fire prevention and protection against burglars and fight, etc. The system arrangements for realizing the village security mainly are: (a) The civil militia organization of township government is set up in Jiezhu village, and usually serve Jiezhu village. (b) Jiezhu village established the security defense team consisting of 6 persons (one person per family) in 2005, which is responsible for looking after property mutually. (c) Select "peace household" every year. Those households that did not fight, violate security regulations or commit crime can be named as "peace household". The township government will hang "peace household" plaque on the door of the household as congratulations.

2.5 Disaster prevention and mitigation

Poverty and returning to poverty caused by disasters happen from time to time in Jiezhu village. Therefore, the government and villagers have implemented direct measures and indirect measures against natural disasters.

2.5.1 The direct disaster prevention and mitigation measures mainly include: (a) Building flood control dam. In 2008, the government spent 1.35 million yuan building the dam around the village which is more than 120 meters high, and effectively realized flood control in flood season. (b) Disasters investigation. The township government established security inspection team, which regularly investigated natural disasters such as debris flow, landslide, collapse, and flood during flood season. Once a disaster happens, the township government will report to the superiors. At the same time, it will organize villagers to relieve the disaster. (c) Potential security dangers testing. The township government security inspection team regularly checks electric safety and building security infrastructure to prevent fire and houses collapse, etc.

2.5.2 Indirect disaster prevention and mitigation measures mainly are: (a) Distribute energy-saving stoves for free. The government researched and produced energy-saving stoves, and distributed them to villagers for free. The use of energy-saving stores can save firewood, reduce deforestation, protect vegetation as well as prevent soil erosion and water loss and mudslides. (b) Cultivate the reasonable slaughter concept toward livestock. In Jiezhu village 10% of families feed livestock in large scale, for example, Zhuzhu's family has more than 260 cattle. Due to the influence of religious ideas, the slaughter keeps low. Zhuzhu' family sold only seven or eight cattle per year several years ago. Snow disaster is the biggest natural disaster facing Jiezhu village, which is prone to cause herdsman family to fall into poverty. A heavy snow at the beginning of 2008, resulted in the freezing 50-60 livestock to death per household for big breeders. Influenced by the snow disaster, the original middle-income herdsman families fell into poverty in 2007. For this reason, the local government has applied a variety of ways to propagate the concept of reasonable slaughter. After the snow disaster, the slaughter rate had risen drastically by the end of 2008, Disasters and the propagation of concept from government increasingly changed the concept of herdsman.

2.6 The exploration of tourism industry

The year of 2000 is the 50th anniversary of the foundation of Ganzi Prefecture. During the celebration for the first time Ganzi Prefecture selected "Kangba Man", and one man in Jiezhu village was selected. In 2003, in the Sichuan, Yunnan and Tibetan Arts Festival, twelve "Kangba Men" from Jiezhu village representing Yajiang county attended the Arts Festival, and showed the special style of "Kangba man". Jiezhu village made its reputation. In 2003, Yajiang county Tourism Administration registered the "Kangba man village" trademark for Jiezhu village. The tourism industry of Jiezhu village started from this year. From 2003 to 2006, there were 15 households reconstructed to be "receptional dwellings" families. These 15 households rebuilt the toilets, perfected the accommodations reception conditions, separated the human from animals; the Tourism Administration carried out culinary training towards these householders, unified the price of eating, lodging and riding a horse. In addition, it has paid more than 400,000 yuan for building the bridle path towards Guogangding, including county financial aid of 90,000 yuan, and the rest invested by villagers in the form of their labors.

From 2003 to 2007, it was the best period in tourism development of Jiezhu village. In 2007, the villagers average income has increased from previous 800 yuan to 1,400 yuan. The "receptional dwellings" households' income was the highest, the average net profits come to 10,000 yuan. Villagers have more access to the outside world through the tourism development.

2.7 Others

In recent years, the local government implemented relocation in order to totally improve the production and living conditions of impoverished population in poverty alleviation and development. Around 100 households have been relocated.

3. Interpretation of the Poverty Reduction Effect of Jiezhu Village

Theoretical analysis holds that the village public products supply can enhance the capability of villagers comprehensively, from reading books and newspapers to ability enhancement in regard to social interaction, and economic activity, participation in community governance, and so on. In terms of the economic activities, the village public products supply will reduce the total cost, including production cost, transportation cost, cost of sales, risk cost and decision-making cost, the natural risk and economic risks in agriculture, so as to improve the efficiency of the production activities; improving rural public product will promote the specialization, scale, commercialization, industrialization, marketization, and sustainable development of agricultural production.

After years' development of village public products supply, the actual capability of Jiezhu villagers has risen drastically. In Jiezhu village, many old men told the author: "We have the best time now. We constructed houses, and the government built the roads, provided electricity, and also issued various kinds of subsidies. Now people don't have to worry about food, clothing and shelters, only care about sideline production and cash income. The price of Chinese caterpillar fungus and matsutake are the hot topics among the villagers in daily conversations." Through the statements from the elderly, we can find the outstanding effect that the government has achieved in poverty reduction by public products supply over the years, namely, the villagers' basic living can be guaranteed. However, low capability is caused by comprehensive factors, and enhancing the capability could not come true overnight. There is still a long way to go.

3.1 The evaluation of the villagers towards the changes

It is an available approach to examine the effect of the supply of public products in terms of poverty reduction according to the changes of the village and the household livings. When surveying in Xi'eLuo county and Jiezhu village, the author investigated more than 20 people and asked the same question as "what's the biggest changes in your family and village during the past ten years?" To sum up their answers, there are three biggest changes in the village: the perfection of transportation conditions, and the upgrading of traffic tools; the abundant power supply accompanied by the popularization of home appliances; the gradual improvement of education conditions, high enrollment rate of school-age students and good employment of part of the village children.

3.1.1 The improvement of transportation conditions

The improvement of transportation conditions has enhanced capability of villagers in Jiezhu village: travel became convenient; the possibility and opportunity of interaction with outside world greatly increased; traffic tools were upgraded, and passenger and cargo traffic became one of the pillar industries; the convenient transportation contributed to the development of the tourism industry.

From villagers' point of view, it's not easy to access highway for every household. Aya told the author: "In the mid 1980s, someone bought two bikes, which was big news for the whole village, and created a mob scene. In 1988, the village had the first tractor, which was still unusual for the whole village, and they gathered and watched. In 1992, a villager bought the first tractor on his own, everyone regarded the family as the rich. In 1993 and 1994, there were 6 and 7 households who bought tractors, some people bought new tractors, other people bought second-hand tractors from Yajiang. Now every family has a tractor, 1~2 motorcycles; there are 3 households who bought big trucks, running a small transportation business, seven families bought cars to drive a taxi, they can earn 300 to 400 yuan each day, or 100 yuan at least."

3.1.2 Abundant electric power supply, popularization of household appliances improved the quality of life

Previously, the village had no electricity, only the District Working Committee generated electricity by generator every night, the power went out at ten o'clock in the evening. In 1986, the public welfare funds and accumulation funds of the entire town from five villages were all invested in construction of hydropower station, and all the villagers volunteered their labour. The power station was finished in that year, every village and every household got access to the electricity. With the abundant power supply, household appliances gradually popularized. The old man named Renze told the author, "In 1984, people only had radios and tape recorders, now 100% of families have TV, tea grinders, mixers, 30% of the families have freezers and washing machines". The popularization of home appliances made life convenient, saved the labor force, and kept visitors staying easily. In addition, the elderly signed with emotion especially, saying that the access to electricity enabled children to read books and do homework more conveniently in the evening. People can watch TV during leisure time, and the cultural life becomes more abundant.

3.1.3 The development of basic education

From the narration of the several old people such as Duojie, Zhuzhu, and Zeren, etc., the development of basic education in Jiezhu village can be concluded as follows: In 1959, the township built a primary school with only 2 or 3 teachers. At that time villagers lived a hard life, and were reluctant to send their children to school. During the political movement in the early 1960s, students were called on to take part in physical labor. The villagers thought it didn't matter whether children went to school or not. In 1965, a student graduated from the village primary school with excellent grade and was employed as accountant in the Township Credit Cooperatives as a government cadre. Everybody admired him very much. The village children then had some learning enthusiasm. During the "Cultural Revolution", the school still ran. After the "Cultural Revolution", some students found jobs after graduation, villagers had some interest in education, but still 50% to 60% of families didn't value education. They asked children to collect Chinese caterpillar fungus and matsutake instead of studying. Now villagers attach great importance to education, the reason is that part of graduates from the primary school went to colleges and universities and got good jobs, and some of them even became government cadres. For example, Tashi is the chief of Immigration Bureau in a county. Liu is the director of the Tourism Bureau in a county. Zhang had been the deputy secretary of the CPC in a county. Zhang's brother is chief of the City Planning Bureau in a county. In addition, there are examples of Jiayang and Fu and so on. There are so many college students and government cadres, which greatly inspired the learning enthusiasm of teenagers. In 2008, three young people in Jiezhu village were admitted to the civil servants and teachers; in 2009, there were seven young people who were admitted to the civil servants and teachers. In the 65 households of Jiezhu village. Six households of seven young people were admitted to civil servants and teachers, which generated great encouragement to the teenagers in Jiezhu village. At the same time,

Jiezhu village made its reputation in nearby county even in the Ganzi Prefecture. People all appreciated Jiezhu village for its high level of education. In 2009, two daughters of the old man named Aya were admitted to teachers. Aya was very happy, and he believed that education would change future. In addition, the school carried out bilingual education of Tibetan and Chinese, making effective achievements in teaching. Especially, the government has taken the Tibetan language as a subject in civil servant examination, which greatly stimulated the adolescents' learning passion.

3.2 The long way towards poverty reduction

Jiezhu village has made great changes by years of hard working. However, there is still a long way towards poverty reduction.

3.2.1 Income situation is still severe

The proportion of impoverished population in Jiezhu village still remains high, thus further income growth is needed to strengthen capability. In terms of the poverty in economic income, there are still 1,233 impoverished population in Xi'eluo township, poverty rate is 38%; the impoverished population in Jiezhu village are 193 people, the poverty rate reached 33%. Alough there are lots of low-income people, the income-increasing sources are very limited.

Neither pure husbandry household nor pure agricultural household exists in Jiezhu village, every family does farm work, herding, and is engaged in sideline work (mainly the collection of Chinese caterpillar fungus and matsutake). The mode of agricultural production in Jiezhu village is one year one harvest season. 85% of the family only have 2 to 5 yaks, 1 to 3 horses. Farming and husbandry production only guarantee the most basic living needs. Most of family cash income are mainly from the collection of Chinese caterpillar fungus and matsutake. The relatively limited income makes it difficult to cope with household spending, especially for the family whose children go to school in other places. The author examined the rough incomes from collection of Chinese caterpillar fungus and matsutake in the family of Gele, Tashi, Ajia, Gequ as follows.

	The income of excavating Chinese caterpillar fungus		The income of collecting matsutake
	the year of 2008	the year of 2009	the year of 2008
Gele's family	more than 10,000 yuan	more than 1,000 yuan	more than 2,000 yuan
Tashi's family	more than 10,000 yuan	more than 8,000 yuan	more than 5,000 yuan
Ajia's family	more than 10,000 yuan	more than 5,000 yuan	more than 1,000 yuan
Gequ's family	more than 3,000 yuan	hundreds of yuan	hundreds of yuan

Note: In 2009, because of less dry caterpillar fungus and lower of price, the income from collection of Chinese caterpillar fungus was fewer. When the author investigated in Jiezhu village no one start matsutake collection, so the income cannot be calculated.

Although the tourism in Jiezhu village is a good way to increase income, there are many problems to be resolved. ① For example, in 2008, influenced by the "3. 14" event in Lhasa, there were no tourists, and the locals failed to make money; in 2009, even in the tourist season of mid-July, there were few visitors. To increase income by developing local tourism still asks for the further improvement of tourist environment and facilities.

In Jiezhu village, there are still some families in extreme poverty, especially the family with children studying in another places. For example, Gequ and his two aunts forms a family, the older aunt is old and frail, and the younger aunt has been disabled since childhood. Their family is the poorest family in the village. Gequ is studying in the second grade in a high school in a Tibetan school. His tuition relies on the support of the civil affairs department. The monthly subsidies is

---

① Real problem mainly includes: The first is the impact of special politics environment in Tibetan areas. The second is insufficient viewing point. The main tourist attraction is the sea of flower in Guogangding in July and August, but the sea of flower belongs to the nature, without plant, it only can keep more than a month. When the flower fades, the sea of flower does not appear, so the viewing point disappears.

hard to guarantee his basic living, Gequ often does part-time jobs to earn living expense during every vacation.

3.2.2　Weaker capability in the process of income growth

During the investigation of the capability of Jiezhu villagers, the author found that: Few young and middle-aged people went out to work; immigrant population were engaged in handicraft in this village as the village lacks craftsmen; business profits were also earned by immigrant population, etc. To a certain extent, it greatly influenced the villagers' income improvement under the circumstance, and also showed the weak capability of the villagers in relatively isolated mountainous villages during the process of income growth.

The reasons that few young and middle-aged people in Jiezhu village went out to work has two interpretation: One is no need to go out. A few villagers believe that the natural condition in Jiezhu village is good, since there are Chinese caterpillar fungus, matsutake, and Chinese herbal medicine for collection, which can achieve self-sufficiency by and large. The villagers spent time in collecting Chinese caterpillar fungus, matsutake, Chinese herbal medicine, and they just waited for collection season at home, instead of going out to work. Second is caused by the comprehensive reasons, and the main reason is the low educational level which resulted in the difficulty in going out to work. Most of the villagers think that, the person aged 30 to 50 have low educational level in Jiezhu village, some of whom can neither speak Chinese nor learn literate, so they will encounter many difficulties if they go out to work, let alone it's not easy to find a job outside.

According to the author's observation, the business profits earned by outsiders highlighted two aspects. First of all, most of the additional value of agriculture and husbandry sideline products are not substantially owned by Jiezhu village residents, and immigrant business men earned this part of profits. The sideline products in Jiezhu village mainly include Chinese caterpillar fungus, matsutake, Chinese herbal medicine and so on; husbandry products including milk dregs, butter, cowhair and wool, etc. Due to the weak commereial awareness of villagers, and lack of the ideas of deep processing, going out to sell and carrying out large-scale operation, most of the existing merchandise trade basically through the traditional marketing approaches, namely, that local villagers directly sell the husbandry and sideline products to the businessman coming around at the prices of initial raw materials. The villagers think that using such marketing methods is not only continuation of the traditional marketing way, but also related to the educational level of villagers. Secondly, the existing four stores are all run by the immigrant Han people in Jiezhu village. The author observed that the products which the four stores are selling covered every aspect of villagers' production and life. During the transactional process of "a sale and a purchase", Jiezhu village residents have lost the profits, benefits and job opportunities.

Jiezhu village is lack of craftsmen in carpentry and Tibetan coloured drawing or pattern, which results that the relevant opportunities of employment and cash income have been completely taken by the outsiders. The Tibetan architectures are of stone-wood structure and wooden furniture highlighted indoors generate great demand for carpentry. The common coloured drawing adornment inside Tibetan buildings, also need more Tibetan painters accordingly. It's a pity that the author found in Jiezhu village that the six carpenters living in the village are all from Ya'an county of Sichuan Province, the two colored drawing craftsmen are from Dege county, there was only one Tibetan colored drawing craftsman in the village who was still on the starting stage.

The weak capability during the process of income growth enlightens us that according to the village public products supply, in addition to providing better basic education, it's urgent to carry out relevant professional skill training (such as carpentry, Tibetan painting) to help build commodity trading platform and strengthen the consciousness of commodities trading, etc.

3.2.3 Expanding the usage of the alternative energy

Jiezhu village is located in artic-alpine region. Therefore, the need of heating materials is great all year round. At present, the main fuel which Jiezhu village residents use is firewood from mountain forest logging. The direct effect of the construction of alternative energy facilities, such as popularity of firedamp pool and solar energy is to save firewood, reduce deforestation, protect forest, prevent natural disasters such as floods and debris flows from the source.

4. Brief Conclusion

First, Sen's research has expanded the connotation of poverty, he presented that poverty not only means less income, but also low capability, and poor living quality, etc. Multi-dimensional poverty commentary lead the construction need of social policy to adopt diversified poverty reduction approaches and focus on capability. The poverty reduction goal is not only raise income, but also struggle to realize the freedoms which people can actually enjoy life and really possess life, to acquire more freedom of action, to obtain more opportunities, to make more choices, namely, have a life which deserves one's cherish.

The ideal pursuance for poverty reduction confronted with the reality of the concentrated special type poverty-stricken areas such as Ganzi Prefecture. Although poverty reduction result has been achieved, it still has a long way to go. The concentrated special type poverty-stricken areas integrated various vulnerabilities: Little and barren land in alpine valleys accompanied by natural disasters, minority relative inhabited or scattered, insufficiency public products supply and weak capability, the less developed economy accompanied by capital and human resource scarcity, fragile industrial development accompanied by huge market risk, and so on. Various fragilities are in urgent need of the poverty reduction "breakthrough" in all directions: From the most basic survival capability of villagers (such as healthy body, ability to read books and newspapers, ability to understand and speak Chinese, etc.) to the productive capability (such as certain planting and breeding techniques, etc.), and then to the developmental capability (such as dealing with market risk, improving the ability of community participation); from infrastructure to industrial development, and then to the ecological and sustainable development in economy, society and culture. At the same time, there is a great necessity to set up demand-oriented poverty relief mechanism, in view of the different groups with different difficulties, not only safeguard the basic survival but also improve the ability to develop by themselves.

Second, the idea of capability from Sen has expanded the village public product connotation. Sam observed poverty on the basis of capability, social right and the quality of life. However, the capability, social right, the quality of life of the villagers are public products. In the research of village public goods supply, currently, the academic circles attach more importance to "tangible community public products" such as infrastructure construction, basic education and medical care. The concept of capability lead the academic research and practical field to pay a further attention to social opportunities, field community participation and social interaction. The practice in Jiezhu village tells us, we can alleviate poverty through the village public products supply. According to the current poverty reduction model, whether the propelling of concentrated continuous development by the whole village or the whole township, we all need to start from personal life to improve the capability of impoverished population. In general, namely, we should focus on "three basic problems" to reduce poverty: Improve the basic production and living conditions, increase the basic quality, and strengthen capability. Focus on the self-development ability of villagers while strengthening their capability including the enhancement of self-confidence, capacities and income.

Third, in concentrated special type poverty-stricken areas, the improvement of impoverished population capability needs the coordination of government, social organizations and villagers. However, compared with other non-concentrated special type poverty-stricken areas, it emphasizes more on the government's responsibility, and even the responsibility of the central government. It not only embodies the actuality in the improvement of impoverished population capability in the concentrated special type poverty-stricken areas, but also embodies the theoretical realization leading by the government①. Just as what the elderly named Zhuzhu has spoken to the author when asked about the biggest feelings about the real life, when the author investigated in Jiezhu village, the old man said, "Individuals are too weak to depend on themselves; there's no use to rely on the God, for the climate is too bad; there's no use to count on the temple, the religious believers go to pray, the unbelievers have nothing to do with that; nothing but the government alone can we rely on. Everything should depend on the government. We will turn to the government for help when we are in trouble such as falling ill. When there are economic difficulties, we can only resort to the government. Apart from that, what can we do?" According to the complexity which is inherently constructed by the concept of capability and the arduousness of poverty reduction practice in Jiezhu village, we may hold that it is self-evident that the public actions which bear the social security function are neither just the isolated national activities, nor alms, or a kind of redistribution of charity. Moreover, it should be the overall social activities participated by the public. However, in reality, the absence of social organizations in the concentrated special type poverty-stricken areas such as Jiezhu village, would restrict poverty reduction process.

Fourth, the concept and practice of poverty reduction are in accordance with nationality. The concentrated special type poverty-stricken areas, are mainly ethnic minority areas. The combination of natural environment, economic mode, national tradition and regional culture on which the people of ethnic areas live formed the nationality of their respective characteristics, which led to big differences in each ethnic traditional cultures and value concepts. The poverty reduction concept we constructed should not be completely against their national

---

① Li Xueping, Wang Zhihan. *The Short Board Effect: Tibet Public Products Supply and Balance of Public Products Supply in Tibet*. Guizhou Social Sciences, 2009, Vol, 2, PP. 101-106.

concepts. Moreover, we need to fully exploit their local and national knowledge to be in line with their national culture ideas. Even if national concept is hard to change in a short time, we also need to take a little longer time; our poverty reduction practice should not be abrupt "external implantation", but should be "internal generation" combined with its soil; the purpose is to achieve "endogenous type" development through the realization of poverty reduction.

References:

[1] Feng Ying. The Evolution of the Definition of Poverty and Thinking about Poverty Issues in China[J]. Economic Research Guide, 2010(8).

[2] Zeng Jingjing. A New Interpretation of Equalization of Basic Public Services [J]. Journal of Nanjing University of Technology(Social Science Edition), 2010(2).

[3] Liu Liu. Rural Public Product Supply and Poverty Alleviation [J]. Journal of Guizhou Provincial Committee Party's School of CPC, 2009(6).

[4] Liu Liu. The Impact of Rural Public Products Supply in Guizhou to Alleviate Poverty Based on the Structural Analysis of Poverty Relief Funds [D]. Guiyang: School of Public Management in Guizhou University, 2008.

[5] Jin Shuanghua. The Public Products Supply and the Income Gap between Urban and Rural Areas[J]. Journal of Dongbei University of Finance and Economics, 2008.

[6] Sui Dangcheng. Research on Rural Public Product Supply Structure[D]. Xianyang: Economics and Management Academy of Northwest A & F University, 2007.

[7] China (Hainan) Reform Development Research Institute Group. The Realization of Human's Comprehensive Development[M]. The Basic Public Service and China's Human Development[M]. Beijing: China's Economic Press, 2008.

[8] Xu Yi. The Defects of Rural Public Product Supply System and Reform Outline[J]. Journal of Anhui Technical Normal College, 2005(2).

[9] Amartya Sen. Development as Freedom[M]. Beijing: Renmin University of China Press, 2002.

(Translated by Ji Xing)

Mechanism Innovation and Enlightenment to the Combination of Post-earthquake Reconstruction with Development-oriented Poverty Reduction for Poverty-stricken Villages in the Earthquake-stricken Wenchuan of Sichuan Province

Xiang Xinghua　Qin Zhimin[*]

Abstract: This thesis expounds the background of combining post-earthquake reconstruction with development-oriented poverty alleviation/anti-poverty of poverty-stricken villages in Wenchuan earthquake-stricken area, introduces the fundamental measures and innovation system of combining post-earthquake reconstruction with development-oriented poverty reduction of poverty-stricken villages in Sichuan Province, and deeply analyzes the achievements and challenges of post-quake reconstruction in poverty-stricken villages of Sichuan Province. In the post-reconstruction period that the poverty-stricken villages have accomplished the basic task of reconstruction, the poverty reduction

---

[*] Xiang Xinghua, a vice dean of Project Management Centre in Poverty Reduction and Immigrant Office of Sichuan Province; Qin Zhimin, a graduate student for a Master's Degree of Sociological Academy in Central China Normal University.

system of Sichuan Province adheres to the principles of the combination of helping the poor and assisting the needy with post-quake reconstruction, development-oriented poverty reduction with special industrial development, and works hard to explore the corresponding innovation mechanism based on applying such successful experience of pilot villages on reconstruction as pilot of post-disaster poverty reduction for poor villages of stricken areas, overall village poverty reduction for poor villages in stricken areas, classified guidance for poverty-stricken villages. At last, this thesis sums up some enlightenments on mechanism innovation for the combination of post-quake reconstruction with development-oriented poverty reduction of the villages in Sichuan Province.

Key words: Post-earthquake Reconstruction of Poverty-stricken Village; Multi-sectors Cooperation; Key Role of Villagers; Comprehensive Development-oriented Poverty Reduction in Villages; Disaster Prevention and Reduction

1. Background: The Combination of Post-earthquake Reconstruction with Development-oriented Poverty Reduction

In China, at 14:28, on May 12, 2008, a great earthquake 8 on the Richter scale occurred in Wenchuan County of Sichuan Province. Wenchuan earthquake caused enormous loss of lives and properties in such affected areas as Sichuan, Gansu and Shannxi Province. The quake-stricken areas were highly overlapped with the poverty-stricken regions. Among the 51 hard-hit counties, there were 41 impoverished counties; the affected areas had 14,337 administrative villages. Poor villages accounted for 4,834 and the affected population reached 2.183 million. In terms of Sichuan Province, among the 39 extremely hard-hit and hard-hit afflicted counties, there were 7 key counties and 24 task counties in the national development-oriented poverty reduction programs. After the earthquake, there were 2,117 poverty-stricken villages hit seriously in severely affected areas. Besides, 399 non-poverty-stricken villages went back to poverty. The total number of poverty-stricken villages increased to 2,516 in the whole province.

1.1 Huge Impact on the development of poverty-stricken village

1.1.1 The scope of poverty extended and impoverished population increased

2,516 poverty-stricken villages among the whole 39 heavily quake-stricken areas were affected at various degrees which covered 150,000 impoverished population, accounting for 21.5 percent of all affected village households. At the same time, these households have suffered serious loss concerning human capital, living facilities and means of production. The poverty incidence rose from 11.68 percent before earthquake to 34.88 percent, and the scope of poverty extended. In addition, among the 100 general quake-stricken counties in Sichuan, there were 14 key counties in the national development-oriented poverty reduction programs and 72 counties had poverty reduction tasks. In these total 86 counties, there were 5,810 poverty-stricken villages. The number of impoverished population climbed from 890,000 to 1,340,000, and poverty incidence ascended from 13.05 percent to 19.64 percent after Wenchuan earthquake.

1.1.2 The serious loss of farmers' capital resulted in further deepening of poverty

In the great earthquake, more than 65 percent of farmers' houses were destroyed; large number of poultry and livestock died and most leading industries were damaged, basic production and means of subsistence were severely destroyed. After the earthquake, farmers could hardly gain property mortage-free support, because the scheduled repayment of debt were delayed in the financial institutions, and then farmer's credit rating was downgraded. Due to casualties of the earthquake, many poverty-stricken villages saw considerable labor loss, and human capital were even seriously lost in some countryside. For example, Xuanping Town of Beichuan County, had 75 households with 250 people and 130 mu arable land before the earthquake. In 2007, per capita income reached about 3,500 yuan, relying on vegetables, pigs, and labour income. However, all people were relocated in the sample rooms of the stricken areas, making a living only by doing temporary work after the earthquake. Nowadays, most households' monthly-income is less than 200 yuan. According to calculation, apart from policy-related increase in income, peasants' per capita net income hadn't regenerated to the pre-quake level by the end of 2010.

1.1.3 Poverty-stricken villages' ecological environment became more fragile

Wenchuan earthquake damaged natural environment such as mountain forests, vegetation, soil and so on, which made poverty-stricken village suffer from more secondary disasters such as landslides, debris flow and dammed lake owing to unstability of geological environment. The soil erosion due to the vegetation destruction, made ecological environment even more vulnerable in poor villages. Farmers' production and living condition were badly destroyed, including infrastructure (roads, drinking water facilities, power supply system and partial cultivated land ruined) and public service facilities (medical rooms and activity rooms). The loss of natural capital and production and living capital badly hindered poor farmers' production, living and development.

1.1.4 More difficulties in poverty reduction work delaying the process

Since 2001, 2,117 poverty-stricken villages in National Planning District of Sichuan Province have gradually launched poverty-alleviation new village project, 1,587 of which had completed New Village Construction Poverty-reduction. The cumulative investment was 1.62 billion yuan. Sichuan Province has input 178 million yuan for industrial development, and built a large number of industries with local characteristics for income growth. Wenchuan earthquake caused destructive devastation to these villages which just gained developing conditions. Production and living infrastructure which has been built through many years of hard work were mostly ruined. The earthquake also wrecked developmental foundation, which was just established by 30 years' of reform and opening up and over 20 years' development-oriented poverty reduction, and prevented Sichuan Province from accomplishing the task for "the Outline of Poverty Reduction and Development of China" as scheduled, so the development process was delayed.

1.2 The Great Significance of Combining Post-earthquake Reconstruction with Development-oriented Poverty Reduction

1.2.1 The explicit claim from the Central Party Committee and the State Council

The Central Party Committee and the State Council put great emphasis on the earthquake relief work and post-earthquake reconstruction in poverty-stricken villages of earthquake-stricken areas. Shortly after the earthquake, General Secretary Hu Jintao put forward: "It is vital to combine New Socialist Countryside Construction with development-oriented poverty alleviation together in quake-stricken regions." On June 21, 2008, Premier Wen Jiabao, stressed when hearing the reports on earthquake relief work of Shannxi and Gangsu Province: "To combine reconstruction with poverty relief work, render further support to the impoverished areas, fundamentally change the production and living condition so as to promote economic and social development in poverty areas." On June 24, 2008, Vice Premier Hui Liangyu made instructions on the report—Report on Holding a National Seminar for Deans of Poverty Reduction Development Office—written by Fan Xiaojian, who was the director of the State Council's Leading Group Office of Poverty Alleviation and Development: "As for the poverty alleviation work in the second half year and in the future, we shall insist on the principles of development-oriented poverty reduction, promote the combination of the rural subsistence allowance system with poverty reduction efforts, push forward the combination of disaster prevention and mitigation with poverty alleviation efforts, carry out the combination of post-earthquake reconstruction with poverty alleviation efforts. "

1.2.2　The pressing aspirations of affected mass in impoverished areas

Earthquake caused mass severe losses of natural capital, physical capital and human capital, and deepen poverty in poverty-stricken villages. It's difficult for villagers to regenerate only by their own strength, and so they urgently looked forward to support and help from the Communist Party of China and the central and local government.

1.2.3　The obligatory responsibility of poverty reduction system

The impoverished population were the objects of poverty reduction development system in poverty-stricken villages. It was a bound duty for poverty reduction system to promote post-earthquake reconstruction in poor villages. The help toward villagers reconstruction not only fully indicated the Party's and the government's concern, but also extensively practiced the People-oriented Scientific Outlook Development.

2. Main Approach and Innovation Mechanism

To successfully combine reconstruction with development-oriented poverty alleviation in poverty-stricken villages, it was essential for the relevant sectors to explore boldly and innovate actively new mechanism on the corresponding work. The poverty alleviation system of Sichuan was mainly from the following aspects.

2.1　Scientific plan, priority pilot and gradual popularization

2.1.1　Launch pilot village-level plan by widespread participation

The poverty reduction development offices at all levels, after identifying the list of pilot village by competition, set up work group of pilot villages, and then went to these villages carried out deep investigation of the disaster situation as well as the demand of quake-stricken villagers via participation. At the same time, through Villagers General Assembly, all villagers discussed the countermeasures of problems, and then initially set up the reconstruction project framework about pilot villages. Moreover, they held a conference participated by related business technicians in local areas to demonstrate the feasibility of project framework on Villagers General Assembly, and then drew up technology standard and units investment estimates of the subproject in line with pilot villages' situation and market prices. The post-quake reconstruction plan of pilot villages was eventually completed.

2.1.2 Rational formulation for project implementation plan

As a big gap between the actual investment and the planning inputs existed in pilot villages, to make full use of the accumulated reconstruction fund. Sichuan Province organized pilot villages on the basis of the investment those who had gotten, so as to optimize utilization of the investment. In line with the input size, Sichuan Province adopted participatory approach, on the basis of village-level plan, launched General Villager Assembly to discuss and determine priority startup including the processes and methods of improving project implementation and management. What's more, Sichuan Province coordinated relevant professional sectors to design the initial plan and budget to meet villagers' most pressing needs and optimize the utilization of the capital.

2.1.3 Gradual popularization for the experience of pilot villages

Based on the successful experience of pilot village-level project in poverty-stricken village, Sichuan Province worked out Formulation Method of Poverty-stricken Village Plan in Sichuan Province and assembled the cadres and masses from 39 severely affected counties and 2,516 poverty-stricken villages(of them, 2,117 were poor villages, 399 were re-poor villages) of 7 municipalities or prefectures in the whole province, through disaster assessment and investigation of post-earthquake reconstruction demand and referring to project technology standard and units investment estimates of pilot villages' planning. At last, Overall Plan of Post-earthquake Reconstruction of Poverty-stricken Villages in Sichuan Province had been fulfilled.

2.2 Resource integration and efficient propulsion

Due to bad condition and fragile anti-disaster capability, more difficulties in reconstruction and more time were required to regenerate for poverty-stricken villages. Consequently, it was hard to accomplish objectives of post-earthquake reconstruction in impoverished villages, just relying on poverty reduction system. Great efforts had been made to broadly mobilize and integrate various social resources in order to effectively promote post-earthquake reconstruction of poor villages.

2.2.1 Fully mobilize resources within the poverty reduction system

Over two years, 52.05 million of the special funds has been directly appropriated by Sichuan Provincial Development-oriented Poverty Reduction Office and Department of Finance, and then 19 million village mutual aid funds have been invested into 126 affected poor villages and 18 counties without counterpart support of city and province. Each village acquired 150,000 yuan of village mutual aid funds to implement pilot project of poverty reduction to develop industries. Besides, they integrated 990 million funds from Kaschin-Beck Disease of Aba Autonomous Prefecture and other program funds such as New Village, labor, industry, village road and methane project for post-earthquake reconstruction of poor villages.

2.2.2 Establish multi-sector cooperation mechanisms

While coordinated by poverty reduction office at all levels and propaganda of pilot villages' demonstrational effect, the post-disaster reconstruction of poor villages got energetic support and assistance from leaders at all levels and relevant functional departments. The related functional sectors provided free technical support and allocated some project funds for pilot villages and other poverty-stricken villages during post-earthquake reconstruction. For example, Guangming village, Jiang County of Sichuan integrated 780,000 yuan for drinking water facilities construction, methane tank, site rectification of centralized settlements, waste collection, road construction and other projects, which were from "Hanshan Village Project" of Water Resources Department (RMB 300,000), methane tank project of Agricultural Energy Resources Office (RMB 100,000) and construction fund of Building Departments(380,000). Baihezhai village, Tongchuan Town of Santai County, according to the Village General Assembly's discussion, rebuilt

2.7 km of village road, 5 km of pathway and benefited 1,850 people of 6 villager groups, rendering the village road construction standard on effective pavement 3.5 km wide, and subgrade 4.5 km wide. The total investment of repairment and maintenance projects—contract through public bidding—were 1.2194 million yuan, in which 930,000 yuan came from Santai County Development-oriented Poverty Reduction Office, Transportation Bureau and Agricultural Machinery Bureau; 170,000 yuan raised by villagers themselves; business owners input 119,400 yuan, 6,000 villagers involved as the labour force. In the course of the project implementation, Li Defu, senior engineer, the dean of Poverty Reduction Office and test supervisor, Li Shangzhi, who ensured the project construction quality by means of filling noise density, compaction test and rebound deflection test. The project completion and check group consisted of Santai County Poverty Reduction Office, the engineers of Agricultural Machinery Bureau and the cadres of Baihe Village. The check showed that the project of repairment and maintenance of the roads were qualified according to the contract.

2.2.3 Give priority to house reconstruction of the poor farmers

Through making a deep investigation and study of the impoverished farmers' condition and their self-renewal ability, the relevant departments of Sichuan Province came up with the advice that reconstruction needs special policy support, owing to the difficulty of reconstruction. In order to guarantee the affected households can move into new houses in time, Provincial government took all opinions into account and determined every impoverished family subsidy standard increased by 4,000 yuan more than that of general affected households. In addition, they had given priority to the poor farmers on distributing special Party fees and social assistant funds. At the same time, they established housing credit guarantee funds for earthquake-affected farmers to support farmers' house reconstruction and ensure the financial demand of the poor farmers' reconstruction in earthquake-stricken regions.

2.2.4 Fully play the subjective role of villagers

As we know, the impoverished population were both the farmers from poverty-stricken villages and victims of the earthquake. It was basis for poor villages' reconstruction to benefit maximum from correctly assessing the situation of the disaster. Obviously, reconstruction couldn't do without villagers' participation so as to solve their most pressing demand. There was no doubt that villagers would fully play a principal role in the course of reconstruction, which was a short-time and harsh task, besides the government's strong support. Consequently, through many ways such as home survey, community investigation, the villagers' general assembly elected management organization of pilot village project and set up the publicity system, and the villagers took an active part in the project process of planning, implementation, management and supervision. These approaches fully mobilized the initiative of participation, and brought into play the subjective role of villagers.

2.2.5 Strengthen capacity-building

a. Strengthen capacity-building for poverty alleviation system

The Office of Provincial Poverty Alleviation Development held many trainings including project planning, spot planning, and project management. After that, every city (prefecture)-level offices also followed the steps. These trainings effectively pushed ahead the implementation of related poverty alleviation work, and enhanced the working ability and level of poverty alleviation groups.

b. Strengthen impoverished farmers' capacity-building

Besides the provincial government holding practical agricultural skill training and labor force transformation training for farmers of pilot villages, the Poverty Alleviation Development Offices of County (District)-level also coordinated relevant professional sectors to carry out forum of rebuilding rural houses, craftsman training, self-build infrastructure construction and agricultural technical training. Significant improvement was appeared to the capacities of self-organization, self-management and self-development of poverty-stricken villagers through various technical and skill trainings. Moreover, villagers were involved into project planning, implementation, management and supervision.

2.2.6 Scientific management

In line with the requirement of Funds Management Approach of Financial Poverty Reduction by the State Council Leading Group Office of Poverty Alleviation and Development, the government strengthened capital and project quality management, implemented earmarked funds system and reimbursement system. Through specifying the duties of relevant professional sectors to ensure the project quality implementation, and the project supervision team selected by plot villages carried out the supervision on the progress and quality of projects. After the first inspection team consisting of village representatives and cadres from township and villages checked up the project qualification, the Poverty Reduction Office of County organized such departments as finance, transportation, water conservancy, construction, agriculture, auditing and supervision to conduct acceptance and inspection. If the project passed the acceptance and inspection, they should report the surplus capital, and submit it to the superior departments for records. At the same time, when poor villages submit implementation programme, they should regard follow-up management scheme as an essential part and implement such scheme after project being accomplished. For example, Qinghe Village of Jingyang District, set up follow-up management scheme of specially-assigned person in charge of village roads and maintenance management group to protect open caisson and ditch.

2.2.7 Establish and implement publicity system and complaints mechanism, receive the supervision from villagers

a. Establish and implement publicity system

Disclose all the content included the organization of project, implementation plan, project tendering, capital utilization and project progress to villagers through available ways for the villagers' supervision; and then hold various meetings related to project management group, fund management group and supervision group regularly to solve problems in the construction process.

b. Establish and implement complaints mechanism

Most poverty alleviation offices of county, township and pilot village committees would confirm special representative to disclose contact information for complaints, accept various supervisions and complaints such as post-disaster reconstruction project selection, engineering management, quality supervision, project progress and funds utilization from whole villagers, for the purpose of mobilizing all villagers' initiative and realizing villagers' such rights as participation, decision-making, management, and supervision.

3. Post-earthquake Reconstruction: Achievements and Challenges

3.1 Achievements of post-earthquake reconstruction in poverty-stricken villages

Thanks to efforts over the past two years, significant achievements had been made in the 2,516 poverty-stricken villages, covering 39 severely quake-affected counties and hard-hit counties in the post-earthquake reconstruction under the strong support of the State Council Leading Group Office of Poverty Alleviation and Development, provincial government and Party Committee. By the end of February, 2011, 1,224 poverty-stricken villages directly organized by Office of Poverty Reduction Development have all started reconstruction work, and 97.5 percent of them have finished the project; the total investment of 3.066 billion yuan has accounted for 98.94% of planning input (3.099 billion yuan), in which 660 million yuan came from Central Fund, making up 98.95% of planning input, and 50.1 million yuan came from provincial fund (poverty reduction inputs) occupying 96.25% of planning finance; and 2.355 billion yuan devoted by social inputs. The target of "To basically complete task of three years within two years" has been fulfilled.

3.1.1 Production and living conditions have been fundamentally regenerated in poverty-stricken villages of impoverished areas

For instance, in poor villages of disaster areas, rural housing repairment and reconstruction almost have been completed; infrastructure such as roads, water conservancy and public service facilities (education, medical treatment and environmental protection) also have gained some recovery and improvement; the governing foundation and decision-making capacity of the village committee has showed great strength; the cadre-mass relations has become closer; the rural social capital has increased obviously. The results of reconstruction laid a good foundation for the masses to live and work happily and increase their income, and created favorable conditions for poverty-stricken villages on new village construction.

3.1.2 Villagers played an intiative role

Through applying the participatory approach to pilot villages on post-earthquake reconstruction, villagers were involved in the project planning implementation, management and supervision, and public opinions and wishes were so fully reflected that it could ensure the villagers' rights including decision-making, participation, supervision and information-reception, greatly mobilizing villagers' participatory enthusiasm. As for the vital interests such as contracted land, private hills and land, villagers followed the requirements of planning and consciously took the whole situation into account and actively devoted money and labor. They did not wait for or just depend on external support, but worked hard and mainly relyed on themselves, fully reflecting the initiative role in the whole work.

3.1.3 Improvement of self-development capacity of poverty-stricken villages

Village officials and villagers enthusiastically participated in project selection, implementation, management and training, through which they improved self-development capacity in terms of dealing with financial affairs, external coordination, project management and independent decision-making. The village committee of Makou Village initiatively communicated with Credit Cooperative Company of the District and won over more than 20,000 yuan loan for each farmer's housing reconstruction; the Muyu Village committee of Qinchuan County initiatively coordinated the national-level poverty alleviation leading enterprise named ChunZhen Industrial Commerce Corporation and regenerated forest, vegetable, and edible mushroom industry with its assistance, and again established professional association. Here, there were 22 vegetable greenhouses, 250 acres of general vegetable, 108 households planting bamboo fungus, agaric, shiitake; in total 70 acres of shiitake, 32,000 bars of fungus linden wood and 38,000 bags of shiitakes were produced in the whole village.

3.1.4 The cadres keep close ties with the masses and promote harmony in community

The branch of village party committee and party members carried forward the spirit on "A Party branch a fortress, A party member a banner". They fully played the exemplary vanguard role of party members by leading publicity and mobilization and actively devoted labor. Therefore effectively made good impression on the masses, and also maintained close ties with the mass. Through participating in the process of project planning, implementation, management, villagers formed a good custom of mutual help which facilitated community harmony. Makou Village of Leezhou District, under the leadership of branch village party committee, carried out the method of "Unified planning and joint construction, unified construction and allocation" to reconstruct the new village. They combined post-earthquake reconstruction with new village construction, ecological regeneration, infrastructure construction and industry development together for unified planning, implementation and evaluation; at concentrated settlement spots of villages, integrated farmers' finance, material and labor and made them benefit from mutual help and joint participation orderly on rural house reconstruction. The community became more harmonious due to the building of an increasing unity and combating strength of the grass-root organizations during post-earthquake reconstruction.

3.2 Existing problems and challenges

The realization of the goal of "to basically complete tasks of three years within two years" indicated that Sichuan's post-earthquake reconstruction fully went into a new period of comprehensive development and enhancement. Obviously, great improvement has been made in the construction work of rural houses and infrastructure of production and living, etc. There's no doubt that new rural houses and breeding houses, broad and flat village roads all show the enormous change in poverty-stricken villages, and some poor villages even exceeded pre-quake developmental level. Nevertheless, in new period, various difficulties and challenges still existed in the development of poverty-stricken villages in Sichuan.

3.2.1 The ecological environment remains still fragile in poverty-stricken villages

As mentioned above, in spite of regenerating production, living condition, several villages changed better than pre-earthquake through reconstruction, but ecological environment remains fragile in poor villages. Wenchuan earthquake destroyed the mountain forests, vegetation, soil and caused secondary disasters such as landslide, collapse and debris flow owing to unstability of local geological environment. Poverty-stricken villages still need a long time to recover their ecological environment.

3.2.2 Farmers' serious liabilities

After post-earthquake reconstruction, most farmers have built new houses and improved the living conditions. However, according to statistics, poor farmers hardly had savings, and rebuilding house cost so much money that each household had about 40,000 yuan liabilities on average (12,000 yuan per person). At the same time, in order to raise funds to build village infrastructure and other public facilities, the masses invested 500 yuan per household (150 yuan per person) and labor investment was 50 per household (15 per person). Compared to pre-earthquake, farmers from poverty-stricken villages universally had deepened liabilities after the earthquake. In addition, some farmers who loaned money from bank or credit union became poorer as a result of reconstructing house in debt.

3.3.3 The phenomenon of poverty-returning and impoverishment caused by earthquake stands out

Earthquake didn't only wreck farmers' agricultural land, sloping land and woodland, but also wiped out their living houses and colony houses, especially in severely affected areas. Heavy casualties made the lives of poor households more difficult. In Sanfeng Village of Qingchuan County, 32 people died, 167 people got hurt and 1,644 mu land got lost in the earthquake. All farmers returned to poverty. In Anyue County which belonged to general quake-stricken areas, 10% of the total impoverished population went back to poverty because of illness and disasters caused by the earthquake.

3.3.4 Production difficulties

Although the total investment plan, which was integrated in the national project of Post-earthquake Reconstruction Programme on Poverty-stricken Villages in the Quake-stricken Wenchuan of Sichuan Province reached 8.998 billion yuan (3.57 million yuan per village), the practical fund used in post-earthquake poverty village reconstruction was just 1,436 billion yuan from Central Fund (570,000 yuan per village) and this led to great shortage in funds. In light of farmers' urgent needs, the Central Fund was mainly put into reconstruction of major infrastructure projects, including village roads, water and drinking facilities and so on in order to meet farmers' aspiration and maximize the use of the limited funds as much as possible. As a result, little investment was allocated to the farmers' production and regeneration. Therefore, they hardly had savings and were even indebted due to low income, house reconstruction as well as raising fund to build village infrastructure and other public welfare facilities. It was very hard for impoverished population to regenerate just relying on themselves under the condition of lacking start-up capital, and hence sustainable development cannot move forward in villages. "Reside in new houses with no money at hand and live a poor life" was a real portray of poverty-stricken villagers in quake-stricken areas.

4. Exploration: The Combination of Post-earthquake Reconstruction with Development-oriented Poverty Alleviation in the New Post-earthquake Period

The various challenges that poverty-stricken villages faced in quake-stricken areas of Sichuan Province showed that poverty alleviation has replaced post-earthquake reconstruction to be the most important initiative of sustainable development in the new post-earthquake period. To effectively resolve the difficulties suffered by impoverished population, and achieve sustainable development of poverty-stricken villages in the new period of post-earthquake reconstruction, Sichuan provincial poverty alleviation system adhered to integrate development-oriented poverty alleviation with reconstruction and adhered to the principle of helping the poor and assisting the needy by developing industries with special characteristics. Provincial Poverty Alleviation and Immigration Bureau implemented the requirements of Guideline of Promoting the Practice of Helping the Poor and Assisting the Needy in the Quake-stricken Wenchuan published by the Provincial Party Committee and General Office, by means of applying successful experiences of post-earthquake reconstruction to poor villages' new mechanisms, which combined pilot project of poverty reduction by developing industries, comprehensive classification of development-oriented poverty alleviation in villages and poverty-stricken villages under the guidance of Helping the poor and Assisting the needy in the quake-stricken areas. The main practical activities were as follows:

4.1 Complete Overall Plan of Post-earthquake Poverty Alleviation for Impoverished Villages in the Quake-stricken Wenchuan of Sichuan Province

To push forward sustainable development of poverty-stricken villages, Sichuan Provincial Poverty Alleviation and Immigration Bureau, continuously strengthened support, through in-depth research, and completed the Overall Plan of Post-earthquake Poverty Alleviation for Poverty-stricken Villages in the Quake-stricken Wenchuan of Sichuan Province (2010—2015) (In the following part it is called Plan for short). The total planned investment was 38.1 billion yuan, and planned areas included 2,516 poor villages consisting of 39 severely affected areas and 5,810 poor villages of 86 general affected counties which had poverty alleviation development task. In July, 2010, the Plan which had been submitted to the State Council by Sichuan Provincial Government, has been adopted into The Twelfth Five-Year Overall Plan for National Economic and Social Development and The Outline for Development-oriented Poverty Alleviation for China's Rural Areas (2011—2020) and has been regarded as a special aspect to support at the fifth and sixth coordination group meeting about post-earthquake reconstruction of the State Council.

4.2 Launch a pilot on poverty alleviation by developing industry in the quake-stricken impoverished villages

The development and revenue growth became main issue in poverty-stricken villages in the period of new post-earthquake reconstruction. In accordance with the requirement of Guideline of Helping the Poor and Assisting the Needy in the Quake-stricken Wenchuan by Provincial Party Committee and Provincial Government, Sichuan Provincial Bureau of Poverty Alleviation and Immigration put into 1 million yuan of financial funds of poverty relief to carry out pilot work in quake-stricken regions, by way of selecting one village in every county of 31 severely affected counties shouldering poverty alleviation task. They have also fully played leveraging role of poverty relief to integrate resources from all sectors and brought about economic development in impoverished villages. Pilot villages relying on their own advantageous resources, considering market-oriented demand and based on fully respect for the villagers' wishes, carefully implemented the project through scientific planning, reasonable layout and laid emphasis on such characteristic industries as farm production and poultry raising to help the poor solve their long-term livelihood in impoverished areas.

So far, cadres from 31 pilot villages and 31 county-level offices of development-oriented poverty alleviation among 7 municipalities (prefectures), have acquired training concerning methods of pilot work and studied the agricultural industrial development experience in developed coastal areas. What's more, they have drawn up village-level planning and implementation programming in pilot villages. The projects have been effectively carried forward.

4.3　Research on the impoverished village classification in the quake-stricken areas

Sichuan Provincial Bureau of Poverty Alleviation and Immigration, with granted fund for projects of technical cooperation between China and Germany, recruited experts from Sichuan Agricultural University to study classification on 2,516 poverty-stricken villages in severely affected areas, through analysis on different types of poor villages and its effectiveness and divergence to decide the developmental strategy. On the basis of classifying poverty-stricken villages, experts determined poverty alleviation inputs and developmental strategy of impoverished villages according to their own geographical and advantageous resources to accomplish guidance on classification. At present, Proposal of Poverty-stricken Villages Classification Research Project in the Quake-stricken Areas has been submitted to the State Council Leading Group Office of Poverty Alleviation and Development and the German Technology Cooperation Company for approval.

4.4　Launch comprehensive development-oriented poverty alleviation of impoverished villages' developmental plan in quake-stricken areas

After cooperating with German Technology Cooperation Company, Sichuan Provincial Bureau of Poverty Alleviation and Immigration as well as the State Council Leading Group Office of Poverty Alleviation and Development utilized project fund, which was related to technological cooperation between China and Germany on post-earthquake reconstruction of poverty-stricken villages, to employ experts from Sichuan Agricultural University to write *Guide Book of Comprehensive Development-oriented Poverty Alleviation in Impoverished*

Villages of Quake-stricken Areas of Sichuan Province (In the following it is called *Guide Book* for short) and guide comprehensive development-oriented poverty alleviation of village-level planning establishment among 31 pilot villages. Based on summary of the above mentioned village-level planning experience, they improved *Guide Book* and then progressively and comprehensively popularized the practice in 2,516 poverty-stricken villages of severely affected areas.

4.5 Progressively implement development-oriented poverty reduction project with eco-compensation benefits

On October 15, 2010, the Fifth Plenary Session of the 17th Central Committee of CPC was held in Beijing and put forward Proposition of Twelfth Five-Year Plan for National Economic and Social Development (In the following it is called Proposition for short). The Proposition explicitly approved the policy—quicken the establishment of a sound ecological compensation mechanism and gradually set up carbon trading market, pointing out the approach of implementing development-oriented poverty alleviation project with the eco-compensation benefits in poor villages of Sichuan quake-stricken areas. Sichuan Provincial Bureau of Poverty Alleviation and Immigration, orderly guided the poverty-stricken villages that were lack of large-scale agricultural production condition towards a green, ecological, environment-friendly, low-carbon path, and made the farmers acquire a stable source of income from establishing comprehensive ecological compensation mechanism and carbon trading market step by step.

5. Enlightenments: The Combination of Development-oriented Poverty Alleviation with Post-earthquake Reconstruction

In nearly three years, Sichuan Province adhered to the combination of post-earthquake reconstruction with development-oriented poverty alleviation, and made remarkable improvements in the living and production conditions of poverty-stricken villages. By means of scientific planning, pilot priority and progressive popularization, the government also emphasized community's participation to fully display the villagers' key role; multi-sectors' cooperation and resource consolidation were performed together with integrated sector funds in reconstruction inputs of impoverished villages. Therefore, to some degree, the above mentioned positive innovative mechanisms which combined development-oriented poverty alleviation with post-earthquake reconstruction have brought about a certain inspiration to sustainable development of poverty-stricken villages in quake-stricken areas.

5.1 Stick to the combination of disaster prevention and mitigation with development-oriented poverty alleviation

Among 592 national-level poverty-stricken counties in China, 70% of them lie in vulnerable eco-regions and suffer enormous harm due to frequent natural disasters. The devastation that Wenchuan earthquake caused on poverty-stricken villages indicated that natural disasters were one of the major causes leading to farmers' poverty. For instance, Diping and Taihong Village of Beichuan County, where arable land was only 0.1 and 0.2 mu per capita after the earthquake. Most farmers' house sites were completely ruined, and relied on government requisition land for rebuilding rural houses alone. It was hard for them to maintain livelihood just based on present arable land to develop production. In 2009, several farmers' reconstruction houses just have been capped, but were quickly buried by "9·24" floods. These farmers had to rebuild houses in debt. Therefore, to ensure sustainable development in impoverished villages of the quake-stricken areas, it was necessary to integrate development-oriented poverty alleviation policy with post-earthquake reconstruction; therefore, relevant departments should profoundly realize and earnestly implement central leaders' important instruction such as "combine development-oriented poverty reduction with post-earthquake reconstruction" as well as "the Scientific Outlook on Development", strengthen responsibilities and sense of urgency. At the same time, it's important to keep in line with natural disasters emergency plans and post-earthquake reconstruction experience summarized in poverty-stricken villages, strengthen communication with relevant sectors and collective investigation, gradually establish mechanisms of mutual supporting emergency in the phase of disaster and post-disaster reconstruction in order to accomplish mutual convergence of funds, policies and measures and maximize their overall benefit.

5.2　Adhere to the combination of post-earthquake reconstruction with comprehensive development-oriented poverty reduction in villages and the whole popularization.

Comprehensive development-oriented poverty reduction in villages and the whole popularization is main approach and vital content in Chinese development-oriented poverty alleviation policies. It is indispensable for poverty relief work to combine comprehensive development-oriented poverty alleviation in villages and the whole popularization with post-quake reconstruction to improve reconstruction and promote sustainable development in poverty-stricken villages. So, it's significant to lay emphasis on establishment of the plan to achieve that purpose.

5.3　Uphold the combination of poverty alleviation policies with rural subsistence allowance system

In accordance with the State Council Leading Group Office of Poverty Alleviation and Development unified deployment, pilot work on "two systems"—development-oriented poverty alleviation and rural subsistence allowance system—were effectively formulated to identify the poor holders and poverty households. For poor holders, the system mainly provided their basic needs; for poverty households, the system improved their basic living and production conditions. By means of pilot projects, a new pattern—"Two Wheel Drive" (rural subsistence allowance system and development-oriented poverty alleviation) — was set up. It could effectively enhance the general level and benefits of development-oriented poverty alleviation in poverty-stricken villages and make impoverished population shake off poverty and become rich as soon as possible.

5.4　Uphold the combination of external help with stimulating internal vigor

The Party Committees and governments of earthquake-stricken

areas endeavored to assist the poor and showed fully respect for their key role. With full enthusiasm and wisdom, they also encouraged farmers to insist on the self-reliance and hard work spirit—Don't wait, don't loosen, don't stop—so as to rebuild a better home with their own hands.

5.5 Uphold the combination of development-oriented poverty reduction with developing characteristic industries

Integrate resources from all sectors so as to bring about economic development in poverty-stricken villages. Depending on their own advantageous resources, market-oriented demand and fully respecting the villagers' wishes, Sichuan poverty alleviation system carefully screened projects through scientific planning and reasonable layout, which were mainly devoted to such characteristic industries as farm production and poultry raising to help the poor solve their long-term livelihood in impoverished areas.

5.6 Uphold the combination of government dominance with social assistance

Great efforts should be devoted to coordinating the Central and Provincial government offices which take the responsibility for poverty relief, and activities—leaders take charge of spot; sectors undertake village; cadres help households, and strive for more support from all sectors for villagers in the quake-stricken areas, and then further mobilize social forces to participate in helping the poor and assisting the needy. In the end, a new pattern of "General poverty alleviation" could be really set up.

(Translated by Wu Dan)

Sustainable Utilization of Natural Resources and the Environmental Risk Management Strategies of Poverty Alleviation and Development

—A Case Study of Sustainable Use of Wild Chinese Herbal Medicine in Daping Village, Pingwu County, Sichuan Province

Deng Weijie[*]

Abstract: For both poor households and poor communities' survival or development, natural capital is the most important livelihood capital. Once the natural resources of livelihood capital, especially the renewable and sustainable utilization can be realized, it could not only provide an effective guarantee for communities' environmental disaster management, but also the basis for communities' sustainable development. The practice of Daping village in Pingwu county, namely Sichuan Province's Community-based sustainable use of wild Chinese herbal medicines provides an effective model of environmental risk management and poverty alleviation and development. The environmental risk management model based on communities' will and management system, not only promotes the sustainable use of natural resources, but also achieves effective community-based environmental risk management

---

[*] Deng Weijie, Associate Professor of College of Tourism of Sichuan Agricultural University.

and it will realize a win-win situation of environment (biodiversity conservation) and community's development eventually.

Key words: Natural Resources; Sustainable Utilization; Environmental Risk; Poverty Alleviation and Development

1. Livelihood Capital Theory and Natural Resources

As we all know, the conventional theory of livelihood capital covers the following five areas:

(1) Natural capital: including land, forest, water, pasture, and other available natural resources;

(2) Financial capital: available and disposable financial resources, such as deposits, loans, remittances, etc. ;

(3) Physical capital: highway, water conservancy facilities, communications and other available infrastructure;

(4) Human capital: including effective amount of labor and labor quality (skill level);

(5) Social capital: the social relations that can be used, including resources from all kinds of group's internal and external social services system (network).

Different groups and individuals have somewhat different or even completely different livelihood capital. For rural areas, especially rural impoverished groups, the differences among the compositions of livelihood capital are very obvious. Different levels of livelihood capital even determine the level of their livelihood's development. For the rural poor, which livelihood capital is relatively more important? The following traditional Chinese proverbs express the essence:

(1) making a living on the agriculature in mountainous areas;

(2) fishing near the sea;

(3) grazing by the grasslands;

(4) doing business in the street;

The above proverbs vividly answered the questions: "What is the farmer? What is the fisherman? What is the herdsman? What is the vendor?" When the farmers lose their land (forests), when the herdsmen lose their ranch, when the fishermen lose the sea, rivers and lakes, they will no longer be farmers, herders or fishermen, and they will lose the livelihood base that maintains farmers, herders and fishermen's attributes, and they will even lose the basic resources that help them develop and get rid of poverty.

2. Environmental Risks

Considering the current situation, in addition to earthquakes, tsunamis and other catastrophes, environmental risks facing poor communities are more commonly pervasive disasters. These include fire, floods, droughts, typhoons, frozens, pests, climate changes, rain, dust storms, snow, mud-rock flows, geological landslides, hurricanes, etc. Different disasters have different risks and different frequency of occurrence. That's why their impact of causing poverty is not the same either. So the corresponding risk management strategies and measures should also be diversified and targeted. For example, we can take engineering measures, such as the construction of breakwaters to prevent tsunamis and typhoons. As for some other environmental disasters, both engineering measures (such as flood control reservoirs, water cellars to prevent drought, etc.) and community capacity-building measures (through the daily management to mitigate disasters effectively) can be taken. These measures include fostering community-based organizations and management systems (pacts), restructuring the types of economic and income diversification, etc. The formation of a stable livelihood capital can not only provide a stable service for the community, including eco-system services and economic services, but also reduce the community's environmental risk, lessen environmental risks' impact, alleviate community's poverty and vulnerability, and promote poor communities' development and anti-poverty. Considering China is a developing country, before engineering measures (artificial construction) that demand huge capital become the main environmental risk management measures, the more realistic and economic measures should be ecological management. That is, taking appropriate natural resource management means to establish and maintain the ecological functions and developmental capabilities of natural resources, to establish and enhance community's ability to defend disasters, to achieve dual purposes—disaster prevention and poverty reduction. Sustainable natural resource management is a very important means and strategy. This will involve community-based follow-up maintenance mechanism, the management mechanism of the sustainable use of community resources, the cultivation of community resources' management and development and so on.

We know that community disasters usually include the following phases or stages:

(1) normal period;
(2) disaster period;
(3) stable period (recovery);
(4) Re-entery to the normal period.

Only when a mechanism based on community's natural resources' management and sustainable use is established in the normal period can we achieve the target of preventing and mitigating disasters and reducing poverty throughout the disaster cycle. In this respect, Daping village in Pingwu county of Sichuan Province has demonstrated very good practical experience.

3. Daping village's Practices: Based on Natural Resources' Sustainable Use

3.1 The basic situation of Daping village

Daping is located in Shuijing town, Pingwu county, Mianyang City, Sichuan Province. As the village is located in the mountains, the locals called it Daping Mountain. It's surrounded by a number of giant panda natural reserves with rich natural resources. According to the 2010 survey, the village has five communities, 84 households and 420 people. It's a mixed community where the Tibetan, the Han and the Hui live together. The village has about 963 acres of land, 8,000 acres of village forest, and its total forest area is up to 15,000 acres. Rich forest resources and the unique geographical environment give birth to very abundant resources of wild herbs. Because of this, just like many high mountains of northern Sichuan, though corn, potatoes, cattle and sheep farming is the basic means of livelihood of the village, collecting Chinese herbal medicine has always been Daping and other neighboring communities' important source of cash income. For some local villagers living by collecting or transporting of Chinese herbal medicines, moving into cities and working there is not their priority of livelihood.

Figure 1 Daping village which lies in the remote mountains

The socio-economic survey shows that wild herbs which are collected most frequently by local villagers are fritillaria, notopterygium, gastrodia elata and cordyceps, etc. The relatively large amount contributes adequately to a household's high income. However, the collection of wild Chinese herbal medicines is in a state of fluctuating acquisition during rush hour or normal period. When the price is lower, less people come to gather; when the price is higher, there are more people gathering. This model is not only unsustainable, but also has resulted in significant damage to the local sustainable livelihood and environmental protection. For this, the World Wildlife Fund (WWF) collaborated with the EU biodiversity projects (ECBP) to carry out a project on sustainable use of wild Chinese herbal medicines. The purpose of the project is to promote better development of local sustainable livelihood and protect the environment (biodiversity) effectively through establishing a mechanism of sustainable use and management of natural resources (wild Chinese herbal medicine).

3.2 The choice of target resources

Daping village' wild herbs are very abundant. However, due to the long-term disordered acquisition, many valuable medicinal herbs (species, as described above) have become extinct. At the same time, different species also face a variety of social and economic pressures, including the collection preferences of the community, market's demand (including higher prices and demand increase), and situation of the market trading norms, etc. Therefore, the choice of the target resource (species) which is based on the sustainable use of natural resources and livelihood development must consider the following factors:

(1) whether the species (state protection level) can be used for statutory;

(2) whether resource ownership is clear (less conflict);

(3) distribution area;

(4) size of biomass that can be collected;

(5) species (self-) renewal capacity;

(6) community recognition (interest);

(7) market potential;

(8) cost-effectiveness;

(9) level of the market competitiveness;

(10) whether communities, particularly disadvantaged groups can benefit from the fair;

(11) the difficulty of establishing sustainable management system;

(12) other factors that must be considered.

In this case, Cordyceps, Gastrodia elata, Fritillaria which are needed intensely by Daping and other surrounding communities are facing the huge market demand for their prices are very high. So, it's very difficult to take standardized collection measures and it's also difficult to get the communities' approval and support if you want them to regulate their collection immediately. After several rounds of consultations, the widely distributed Sphenanthera with better biomass was determined by all parties to be the target species for the project because of its easy collection, low level of protection and huge market potential. They try to take sustainable measures on its usage and management.

Figure 2   Daping village's sphenanthera

3.3   The cultivation of community management organization

After repeated consultation with the villagers, the project's facilitator recognized that the community organizations of wild Chinese herbal medicines' sustainable use and management should be immediately established in the rural areas. Multiple options are presented.

3.3.1   The villagers rent land to the businessmen who live by selling medicinal herbs voluntarily and get rent fee in accordance with the contract; they may also earn money from the businessmen who manage the land independently. The lease is determined by the two sides. The contractual relationship is the relationship between individual farmers and the businessmen, so it is not necessary to establish associations. Now there have been about 30 households in the village renting land to the businessmen in accordance with the 30 years' land use period. This approach is characterized by simple operation so that the villagers do not have to worry about production and marketing. Instead they can receive rent on a regular basis. On the other hand, the drawback is that the villagers' income is not high. At the same time, they are lack of correlation with objectives of the project and have no effect on improving the collection of wild medicinal herbs.

3.3.2 Medicinal-herbs-collecting farmers set up a herbal producers' association and the members exclusively are all villagers of Daping. The association manages it itself and has no other relationship with the businessmen apart from the products trading relationship. The disadvantage of this approach is likely to weaken the community's management. The members may be fragmented easily, because they have to carry out sales activities on their own, thus leaving a superficial existence of the association. Contracts with the businessmen are difficult to be fulfilled, especially when the price is volatile.

3.3.3 The medicinal herbs growers and related parties (such as medicinal herb businessmen, government services, etc.) set up Pingwu Shuijing Chinese Herbal Medicine Manufacturers' Association voluntarily. Members include medicinal herbs growers (villagers and contractors), medicinal herb businessmen, technological service providers (such as town government forestry technical service centers, etc.). The association signs united contract with all parties and then breaks down the production and marketing programs and provides a unified seeds, technologies and marketing services for its members. The members manage the land independently in accordance with requirements of the association and enter the market through the association in accordance with the contract; members shall not sell and can not change the production plan (for example, acreage, location and types of medicinal herbs, etc.) arbitrarily. One of the advantages lies in the risk-sharing. Specifically, this approach brings together the growers, distributors and service providers together to face the risks of production and marketing. Herb growers can reduce the fear of a bad market, while the dealers can reduce concerns that herb growers do not carry out production plans according to the market. The members share profit and loss of Chinese herbal medicine's production and management together.

3.3.4 The medicinal herb growers and herb businessmen set up Daping Village Chinese Herbal Medicine Production and Business Cooperatives and it operates in accordance with the "People's Republic of Farmer Cooperatives Act". This approach involves funding, the proportion of non-participants in the villagers, setting up members' accounts, and setting up the board of directors, supervisors and other requirements, so its operating procedures and requirements are more complex. However, due to the lack of trust among the villagers and between villagers and medicinal herbs dealers, and the producers' (collectors') lack of confidence in institutions and markets, thus there has not been ideal conditions for this option.

3.3.5 Villagers set up Shuijing Chinese Herbal Medicine Production and Management Company together with medicinal herbs businessmen. Both the villagers with land management rights and the businessmen who possess funds and technology can become shareholders. The villagers can not only get bonus according to their shares, but also have the opportunity to work in the company to obtain additional income. The disadvantage of this option is that it will be difficult for small-scale farmers to interfere with the operation of the company and it is hard to improve the community's ability. At the same time, this approach can not guarantee that both production management and biodiversity conservation can achieve the desired results.

3.3.6 Finally, Daping village, Shuijing town government and WWF reached a consensus, deciding to set up the Association of Chinese Herbal Medicine Production and Business. If conditions permit, in the future the association can be converted to cooperatives or corporating form. On October 19, 2008, Pingwu Shuijing Chinese Herbal Medicine Production and Business Association was formally established. All members of the association were all volunteer membership. The Shuijing Town Government also provided an office which was close to the medicine market for the association.

Figure 3　The office of Daping village Chinese Herbal Medicine Production and Business Association

3.4　The community-based management system and capacity building

According to Pingwu Shuijing Chinese Herbal Medicine Production and Business Association's work scheme, the sustainable collection and use of wild sphenanthera was the first project activity of the association. For this, the project provided a range of capacity building measures and other support services for the association, including assisting Daping village in developing scientific collection methods (guide) and processing methods (guide) of sphenanthera and the village's (wild schisandra) sustainable collection management (pacts), association's management charter, etc. And it also helped Daping assembly elect members of the village's natural resource management team, and identified the resource management team's responsibilities, etc., to regulate the operation and management of the association.

Through a year's operation, the system construction and operation capacity building of the association had achieved remarkable results. Members of the association increased to more than 80 households from the original 60 households, which means the whole villagers had become members of the association. Meanwhile, with the help of the project, the association signed a purchase and sale contract with a commercial company from United States. In September, 2009, as the first time to meet the scientific process of sustainable acquisition and processing requirements, 500 kilograms of wild schisandra of high-quality were successfully exported to the United States. It was not only recognized by the US company, but also received another 10 tons new order of sphenanthera in 2010. As the Daping village's collection of wild sphenanthera has received US companies' approval in line with high quality and sustainable use and protection, the producing area's sales price has risen from 4-6 yuan per kilogram in the past to 16-20 yuan per kilogram in 2009, and in 2010, its price was as high as 24 yuan per kilogram. The Community has fully endorsed the concept of "effective protection, sustainable use, high return" and complyed with the pacts of sustainable collection. Safeguarding sustainability of the community resources has become the consensus and conscious action of Daping villagers.

Figure 4 The pacts of Daping village in Daping at the entrance of the village

At the same time, villagers in Daping and other key stakeholders aware that as the association does not have the right to operate independently, it can not sign commercial contracts directly with other companies and also does not have a bank account to receive statutory transfer, which limits the further cooperation with the United States' companies. This leads to the inevitable establishment of an independent business cooperative or company. To meet this need, in October 2009, the Association Management Committee held a plenary General Assembly. In this meeting, they not only reelected the Management Committee members, but also identified the establishment of Chinese Herbal Medicines Cooperatives. All members of the association who would like to join in cooperatives can apply for attendance. In December 2009, the project organized association members and cooperative activists to go out to visit the establishment and operation of other farmers' commercial cooperatives. In May 2010, Pingwu Shuijing Herbal Medicine Planting Professional Cooperatives was established. This has further strengthened the village's natural resources' management and sustainable use which is based on community's organization and management mechanism.

Figure 5 The office of Daping village's Herbal Medicine Planting Professional Cooperatives

It should be noted that Daping village has terminated the establishment and traditional operation of cooperatives and entered market-oriented operation phase and started to get independence from the project. The cooperative's office rent, and the daily operating costs (water and electricity, etc.) were all paid by themselves. Part of the Chinese herbal medicine processing equipment (dryer) that the project had provided still operate in full accordance with the market. Monitoring shows that the phenomena of cutting down trees, cutting vines (to facilitate collection), and early acquisition (scientific acquisition time should be at the end of August to early September each year) which appears in collecting Schisandra in the past no longer exists now. In 2010, through the newly established cooperative, sphenanthera's sustainable collection range expanded from single Daping village to 13 villages of three townships. At the same time, Daping village expanded the model of sustainable collection from sphenanthera to Paris polyphylla, rhubarb, jade radix angelicae pubescentis, etc. In early September 2010, several US companies' bosses conducted field visits to the village. They highly recognized Daping village's pro-poor development and environmental protection practices and the effectiveness by way of promoting the sustainable use of Chinese herbal medicines and the two sides signed long-term strategic cooperation contracts based on this model. This again enhances Daping village's confidence of sustainable management and use of wild medicinal herbs.

4. Inspiration from Daping Village's Practice

The process of Daping villagers' collection of wild Chinese herbal medicines, from the simple living demand-driven to the development for sustainable livelihoods and ecological environmental protection drive, from disordered individual behaviors gradually to become an orderly community-based organization and management mechanism, shows that the sustainable use of natural resources is not only one of the strategies of sustainable pro-poor development for those similar communities, but also a strategy of effective community-based environmental risk management. The achievement and maintenance of the process not only strengthened the natural capital of livelihood capital of the community, but also strengthened the ability of disaster prevention and mitigation and greatly eased the environmental risk's sustainability and impact on the community's livelihoods. Some communities are located in the hilly areas where forests, landslides and floods are not new to the villagers, now this situation has been eased significantly. The Wenchuan earthquake didn't have a serious impact on Daping village through geological landslides and the subsequent floods. Although external support project came to an end in March 2011, the villagers still continue to carry out related work under the guidance of the association and cooperatives. It also shows highly recognition of the community for this model.

Daping village's sustainable use of natural resources and ecological environment protection model has the following revelation to other similar communities in sustainable development and environmental risks' management:

(1) Revelation 1: The community's natural resources with livelihood capital properties must have controllable legal relationship, and the ownership must be clear;

(2) Revelation 2: Communities should have legal rights and opportunities to manage the sustainable natural resources and implement specific management according to effective pacts;

(3) Revelation 3: Communities must have a highly recognition of the vision that fulfilling the target of poverty alleviation and environmental risk management should stick to sustainable natural resource management;

(4) Revelation 4: Market problems should be solved through the market mechanism at last. Promoting the conversion of communities' roles consciously should shift gradually from the simple users of resources to the protectors;

(5) Revelation 5: Credibility of community organizations and operational capabilities including market management capability should be established. Therefore, the establishment of community development and management organization is the fundamental way;

(6) Revelation 6: Establishing an integrated service system for the community.

In this case, the community can not only establish and maintain sustainable livelihoods capital, but also can implement effective community environmental risk management, reduce environmental risk of community livelihoods and promote the sustainable development of community livelihoods.

Thus, the community-based effective mechanism for natural resources' sustainable management is one of the effective strategies of poverty alleviation and environmental risk's management.

(Translated by Cai Zhihai)

A Cost-benefit Analysis of Disaster Risk Reduction

Christoph I. Lang *

Abstract: Because of an increasing number of natural disasters, DRR (Disaster Risk Reduction) becomes more and more important, especially for developing countries. since it's difficult to reveal the benefits of DRR, DRR is often neglected in developing countries. Developing disaster insurance could improve the cost-benefit analysis of DRR, and help developing countries adopt DRR measures.

Key words: Disaster Risk Reduction (DRR); Vulnerability; Cost-Benefit Analysis (CBA); Insurances

A Cost-Benefit Analysis (CBA) is often required as a basis for creating or reviewing a public task, since taxpayers want their money to be used efficiently and effectively. This also can be applied in the process of formulating measures of Disaster Risk Reduction (DRR). DRR is not only crucial for the protection of population and infrastructure, but also a fundamental task of any government (International Strategy for Disaster Reduction [ISDR], 2005).

DRR is a long-term commitment, with often small immediate and obvious benefits. It may therefore be difficult to get approval for the necessary budget and personnel (The World Bank and The United Nations [WB/UN], 2010). Since DRR is strongly linked with

---

* Christoph I. Lang, officer of Swiss Embassy in China.

development. Either developed countries or developing countries to which China still belongs adopt DRR measures to do so in different ways.

DRR is often neglected in developing countries, though they suffer a lot from disasters, and are usually unable to cope with the consequences. Without counter-measures, this may lead to perpetuate fragility. Therefore, sustainable development becomes very difficult to move forward (Department for International Development [DFID], 2005). This paper will outline what could be done to highlight the benefits of DRR, and how CBA can be used in support.

1. Risk, Hazard, Vulnerability and Natural Disasters

Risk is defined here as the product of hazard and vulnerability. Sometimes exposure and coping capacity can be seen as aspects of vulnerability (ISDR, 2009, pp. 25, 30; WB/UN, 2009, p. 25).

A hazard is a "dangerous phenomenon, substance, human activity or condition that may cause loss of life, injury or other health impacts, property damage, loss of livelihoods and services, social and economic disruption, or environmental damage" (ISDR, 2009, p. 17). A natural hazard(only these kinds of hazards are dealt with in this paper) is a "natural process or phenomenon that may cause loss of life, injury or other health impacts, property damage, loss of livelihoods and services, social and economic disruption, or environmental damage" (ISDR, 2009, p. 20). However, it is often difficult to define whether a hazard is indeed "natural", or whether it is—at least partly—man-caused. The former UN Secretary General Annan questioned the term "natural" disaster in 1999. To make the difference is particularly difficult and controversial when it comes to hazards stemming from climate changes (WB/UN, 2010).

Vulnerability, the other risk factor, is described as the "characteristics and circumstances of a community, system or asset that make it susceptible to the damaging effects of a hazard" (ISDR, 2009, p. 30). A hazard combining with vulnerability under certain circumstances can become a disaster. A disaster could be described as a risk which causes "serious disruption of the functioning of a community or a society involving widespread human, material, economic or environmental losses and impacts, which exceeds the ability of the affected community or society to cope with its own resources" (ISDR, 2009, p. 9).

2. Impact of Natural Disasters Depending on Level of Development

a. Changing vulnerability

Statistics show there is a generally increasing number of natural disasters (Swiss Re, 2011; WB/UN, 2010). These disasters affect people differently in countries with a high level of development as compared to countries with a low level of development, and they affect different strata of a society within a country differently, with the poorest often suffering most (WB/UN, 2010).

Hazards—as one disaster component—like tornados and volcanic eruptions, occur naturally in some parts of the world, but rarely in others. Natural hazards, by definition, are not caused by human behaviors and therefore are not decided by the level of development. In contrast to natural hazards, the vulnerability component of disasters is related to the level of development. Under-development is by itself regarded as a vulnerability factor (ISDR, 2005).

Generally speaking, a low level of development means high vulnerability, although not all developing countries are affected in the same way, and there are also differences within one country (Annan, 1999). So, while a high level of development usually means more capacity and less vulnerability decline, a low level of development can also make higher vnlnerability(Rego & Roy, 2007). Vulnerabilities are related to changing demographics, technological and socio-economic conditions, unplanned urbanization, development within high-risk zones, environmental degradation, climate variability, climate changes, geological hazards, competition for scarce resources, and the impact of epidemics such as HIV/AIDS (Annan, 1999; ISDR, 2005).

b. Material loss and human casualties

According to insurance statistics, most material loss due to natural disasters occurs in highly developed countries, with relatively few casualties (Swiss Re, 2011). The reasons seem relatively simple:

• Compared with developing countries, there are more sophisticated and expensive assets around in developed countries;

• There is less vulnerability, with coping mechanism in place to avoid casualties in developed countries.

The situation is almost the opposite in developing countries, with relatively less material loss—because there are fewer assets—but with larger human casualties of disasters—because of higher vulnerability. However, there are more reasons for the difference in statistics between developed and developing countries, which include:

• The issue of taking (avoided) loss of lives and injuries into account in the CBA;

• The lack of insurance institutions.

3. DRR in Developed Countries

Protecting the safety of the population is considered to be a fundamental task of any government. Such protection against disasters can be achieved with the support of DRR programs (ISDR, 2005). In Switzerland, for example, the risks emanating from natural hazards are managed in an integrated approach developed mainly by the National Platform for Natural Hazards (Planat). The approach is overarching including natural and technical science as well as crisis management. It conforms with the principles of sustainability and tries to involve all stakeholders (Planat, 2005).

The risk management concept includes:

• An analysis of hazards and the economic, social and physical vulnerability;

• An evaluation of how much risk a particular society or community is able and willing to carry (including economic aspects);

• A plan for mitigation, preparation, response, and recovery measures.

One important aspect of risk management is the fact that risk can be dealt with by insurances; this is, however, not always available in developing countries, as will be explained below. In Step (b) which has just been mentioned, the people concerned can decide to insure part of their risk. In Switzerland, insurance for buildings against damages from natural disasters (though not always including earthquakes, as this is considered to be too expensive) is even compulsory (Planat, 2005). Also in Step (b), it was mentioned that economic aspects must be included in the evaluation. This leads to the CBA.

4. Analyzing Costs and Benefits of DRR

The decision of any legitimate government to become active in a particular field, or how to choose from different options—not only in DRR—should be based on a CBA. This is because a government is expected to spend tax money—or any other income for that matter—in a reasonable and comprehensible way. A CBA is a technique of comparing all the costs (both tangible and intangible) of a particular course of action with the expected resulting benefits. Whether done in the same overarching risk management approach like in Switzerland, or differently: DRR measures must make sense economically, they should be economically bearable and sustainable (ISDR, 2005; Planat, 2005).

Assessing risks is difficult, as is determining the benefits of DRR. It is even more difficult to get the necessary fund to make a CBA. Because "the benefits of preparation are invisible in the short run, but only after a disaster has occurred. In times of financial shortage, programs whose benefits cannot clearly be demonstrated get short shrift on the fist of budgetary priorities" (Auf der Heide, 1989; similar: WB/UN, 2010). Sometimes when it is possible to do a study and it will show a positive CBA, there still may be no funds eventually because of other commitments and priorities.

5. DRR in Developing Countries

Generally speaking, vulnerability is higher in developing countries compared to developed countries, and higher among the poor than the rich within a country. As mentioned before, disasters in developing countries often claim a large number of human casualties. At the same time, in some developing countries there are only some very basic DRR programs (ISDR, 2005). Because of the financial shortage just has mentioned, it may be even more difficult in these countries, with generally tighter budgets, to be able to make a CBA. Given enough funds, there are still another two major issues which may make a CBA seem negative:

· To avoid loss of lives and injuries is mostly not taken into account in a CBA; considering the "Value of a Statistical Life" (VSL) approach, it remains a complicate issue, for it involves ethical and technical difficulties (Provention Consortium, n. d. ; WB/UN, 2010, p. 117);

· There is often no insurance available or people cannot afford it (ISDR, 2005). Therefore, uninsured loss may often not be reported (Lynch, 2004).

6. DRR as Development Issue

Under these circumstances, the benefits of DRR programs may be difficult to prove and they may be neglected, however, as was mentioned, protection of its population from disasters is a fundamental task for every nation, which should go hand in hand with development. In the year of 2000, the United Nations General Assembly established the Millennium Development Goals (MDG) as overall framework of development. DRR is not considered as a separate goal. It nevertheless is an important issue, as disasters often leave governments no choice other than to channel funds for long-term development efforts to cover high cost of dealing with the consequences of natural disasters (Rego & Roy, 2007). If the UN and its member countries want to achieve the

MDG, they should take DRR into account. Development workers and disaster planners share the common interest and one effective way to achieve the common goals is to emphasize DRR in development, particularly in achieving MDG (Benson, Twigg, & Rossetto, 2007; Rego & Roy, 2007).

a. Taking into account avoided losses of life in the CBA

As was mentioned before, loss of lives does not always appear in disaster statistics, so it may be difficult to establish the VSL; hence the CBA is less likely to be positive for DRR. In order to correct this, the number of deaths (and injuries) should be reflected in statistics of damages, so the information would be more complete; to omit it may be sufficient to account for the loss of the private industry, but not for the economic loss of the society as a whole (Suche nach der richtigen Bewertung von Naturkatastrophen, 2006; WB/UN, 2010, p. 116)

If this was done, the ranking of the total damage to an economy would be greatly different. This is very important for each affected country to get the priorities right, and it also must be taken into consideration when deciding about developmental projects in the framework of bilateral or multilateral aiding cooperation. The higher the VSL is set, and the more people are affected by a disaster, the more necessary it would be to review the ranking of disasters. This would often more likely lead to a shift in the CBA to be in favor of DRR measures.

b. Role of insurances

Since uninsured loss may go unnoticed, a CBA more favorable towards DRR could be achieved if there was a bigger role of insurance institutions. Insurances could contribute to a more comprehensive report of loss of lives and injuries. The market of insurances in developing countries, however, is quite small (ISDR, 2005; WB/UN, 2010). Institutions like disaster insurance and risk insurance are only available in a limited way. Whatever risk a community meets (according to risk management process described above) it often has to bear it itself, without having the option of transferring that risk to an insurance. However, even if there are insurances in the developing countries, it is said that most people are too poor to be able to afford them.

Insurances can contribute to better coverage themselves, increasing their business without being exposed to too much risk. There are initiatives that insurances work together with communities. These communities get reward from the insurance company by implementing DRR measures, thereby reducing the risk of harm to themselves and the financial risk to the company (Schweizerische Mobiliar Versicherungsgesellschaft, 2007).

This is in line with the spirit of what the "Hyogo Framework for Action 2005—2015" calls for, that is, to develop "partnerships to implement schemes that spread out risks, reduce insurance premiums, expand insurance coverage and thereby increase the fund for post-disaster reconstruction and rehabilitation including both public and private partnership. Promote an environment that encourages a culture of insurance in developing countries" (ISDR, 2005, p. 19).

7. Conclusion

Because of an increasing number of natural disasters, which often cause large casualties and severe damage, DRR becomes more and more important. This is true especially in developing countries with a high level of vulnerability where disasters can inhibit development. Nevertheless, it is often difficult to get necessary funds to take DRR measures. Mainstreaming DRR in development assistance is considered to be the best way.

Due to the specific situation in developing countries, a CBA may often not be in favor of DRR measures. A more balanced CBA would make such measures more attractive so as to gain budget plans as well as donors. The CBA could be improved by taking into account the loss of deaths and injuries. Insurances could also play an important role in the process by adapting their report and supporting DRR measures in general.

References:

[1] Annan, K. A. (1999). An Increasing Vulnerability to Natural Disasters. International Herald Tribune. Retrieved April 12, 2011, from http://www.un.org/News/ossg/sg/stories/articleFull.asp? TID=34&Type=Article. Auf der Heide, E. (1989). Disaster Response: Principles of Preparation and Coordination.

[2] The Apathy Factor. Retrieved April 12, 2011, from http://orgmail2.coe-dmha.org/dr/DisasterResponse.nsf/section/02? opendocument.

[3] Benson, C., Twigg, J., & Rossetto, T. (2007). Tools for Mainstreaming Disaster Risk Reduction: Guidance Notes for Development Organisations. Retrieved May 20, 2008, from http://www.proventionconsortium.org/themes/default/pdfs/tools_for_mainstreaming_DRR.pdf.

[4] Department for International Development (DFID). (2005). Disaster risk reduction: A development concern. Retrieved May 27, 2008, from http://www.dfid.gov.uk/Pubs/files/drr-scoping-study.pdf.

[5] International Strategy for Disaster Reduction (ISDR). (2005). Hyogo Framework for Action 2005-2015: Building the Resilience of Nations and Communities to Disasters. Retrieved May 15, 2008, from http://www.unisdr.org/wcdr/intergover/official-doc/L-docs/Hyogo-framework-for-action-english.pdf.

[6] International Strategy for Disaster Reduction (ISDR). (2009). 2009 UNISDR Terminology on Disaster Risk Reduction. Retrieved April 10, 2011, from http://www.unisdr.org/eng/terminology/UNISDR-Terminology-English.pdf.

[7] Lynch, D. L. (2004). What do Forest Fires Really Cost? [J]. Journal of Forestry, 102(6): 42-49.

[8] National Platform for Natural Hazards (Planat). (2005). Strategie Naturgefahren Schweiz [Strategy Natural Hazards in Switzerland]. Retrieved April 10, 2011, from http://www.planat.ch/index.php? userhash=54445209&navID=1030&l=e.

[9] Provention Consortium. Disaster Risk Reduction and Cost-benefit Analysis. Retrieved April 9, 2011, from http://www.proventionconsortium.org/? pageid=26.

[10] Rego, L., & Roy, A. S., (2007). Mainstreaming Disaster Risk Reduction into Development Policy, Planning and Implementation. Retrieved May 20, 2008, from http://www.adb.org/Documents/Events/2007/Small-Group-Workshop/Paper-Rego.pdf.

[11] Schweizerische Mobiliar Versicherungsgesellschaft. (2007). Prävention von

Naturgefahren [Prevention of Natural Hazards]. Retrieved May 28, 2008, from http://www. mobi. ch/mobiliar/live/diemobiliar/engagement/m-409000/m-409001. html.

[12] Suche nach der "richtigen" Bewertung von Naturkatastrophen [Looking for the "correct" evaluation of natural disasters]. (2006). Neue Zürcher Zeitung. Retrieved April 10, 2011, from http://www. nzz. ch/2006/02/25/th/articleDL0JB. html.

[13] Swiss Re. (2011). Natural catastrophes and man-made disasters in 2010: A year of devastating and costly events. Retrieved April 10, 2011, from http:// media. swissre. com/documents/sigma1_2011_en. pdf.

[14] The World Bank and The United Nations (WB/UN). (2010). Natural Hazards, Unnatural Disasters: The Economics of Effective Prevention. Washington, DC, USA. Retrieved April 12, 2011, from http://www. gfdrr. org/gfdrr/NHUD-home.

[15] United Nations General Assembly. (2000). United Nations Millennium Declaration. Retrieved May 20, 2008, from http://www. un. org/millennium/ declaration/ares552e. htm.

DFID's Work & Experience on Disaster Risk Management and Poverty Reduction

DFID

Abstract: When Wenchuan earthquake happened, DFID (Department for International Development) began to focus on integrating disaster reduction into its poverty reduction activities in China, and set up a Technical Assistance Facility for Wenchuan Earthquake Post-Disaster Recovery Programme. The purpose of the Technical Assistance Facility was to help make the recovery more participatory and responsive to the needs of the poorest and most vulnerable. Working with Chinese and international agencies on Wenchuan earthquake post-reconstruction not only increased China's awareness of international examples and ideas, but also increased DFID's awareness of the importance of disaster reduction to poverty reduction, and the significance of sharing China's experience with other countries affected by natural disasters in particular the developing countries.

Key words: Post-disaster Recovery and Reconstruction; Poverty Reduction; Disaster Reduction; Technical Assistance Facility

DFID as the UK governmental developmental agency has worked with the Chinese government on poverty reduction and achieving MDGs for more than a decade. Its bilateral programmes focus on health, education and water & sanitation sectors. Wenchuan earthquake brought a change to DFID's cooperation with the Chinese partners. It made DFID integrate disaster reduction into its poverty reduction activities in China.

When the devastating earthquake happened, DFID contributed £2.2 million to buy food, blankets and tents as immediate relief measures, and reallocated some ongoing programme funds to support the earthquake-affected areas. More importantly, DFID specially set up a Technical Assistance Facility for Wenchuan Earthquake Post-Disaster Recovery to support the Chinese central government and the provincial governments of Gansu, Shaanxi, Sichuan and Yunnan in their immediate post-disaster recovery and reconstruction planning.

The purpose of the Technical Assistance Facility was to help make the recovery more participatory and responsive to the needs of the poorest and most vulnerable. This assistance took the form of seven sub-projects, which covered emergency management at national and provincial level, low carbon reconstruction at municipality and county level, environmental planning at county and village level, health response to disasters, mental health support at the local level, spinal injury rehabilitation, and rural water supply reconstruction planning. These were undertaken through partnerships between leading Chinese institutions and high-profile international counterpart organisations. This combination has enabled the quake-stricken areas to have access to international best practice, the opportunity to test the applicability of these concepts to the situation in rural China, and access to policy-makers at national and provincial levels so that outcomes and recommendations can be used to make policies.

Among the seven sub-projects, DFID worked with UNDP(United Nations Development Programme), State Council leading group on poverty reduction and the ministry of environmental protection on improving environmental planning and natural resources management. Its outputs include the guidance document of "Environmental Risk Reduction in Post Earthquake Reconstruction" aiming to be adopted by community members and local level governments.

The facility was deliberately designed to be wide-ranging and ambitious so that there would be enough freedom to implement activities which were considered as the most important as the situation developed. In addition, the collaborative partnership between national, local and international organisations helped establish the long-term relationships necessary for the effective transfer of knowledge, as well as providing a framework within which international experts could work effectively. International response to disasters in many countries focuses on the immediate relief programmes. This Technical Assistance Facility demonstrated the importance of a parallel but small-scale programmes of providing advice on the wider and longer-term issues.

Working with Chinese and international agencies on Wenchuan earthquake post-reconstruction also increased DFID's awareness of the importance of disaster reduction to poverty reduction, and the significance of sharing China's experience with other countries affected by natural disasters in particular the developing countries. Although DFID had formally finished its bilateral cooperation with China by March, 2011, DFID will continue to work with China through partnership for global public goods, and the south-south cooperation will be a major element of the new programme. It is very likely that DFID will work with relevant Chinese partners and other countries to promote knowledge/technology transfer on community-based disaster risk reduction, and facilitate joint study on how to mainstream disaster risk reduction into local and national poverty reduction and sustainable development strategies. In future, DFID will work in partnership with China for a multi-win situation from the global development perspective.

Disaster Risk Management and Poverty Reduction: International Experiences

Sanny R. Jegillos[①]

Abstract: Poverty and vulnerability to natural hazards are closely linked and mutually reinforced. Marking off intensive risk and extensive risk would improve understanding of linkage of disaster and poverty. Social protection should be integrated with disaster risk management. In the practice of poverty reduction and development, disaster risk reduction should be routinized and mainstreamed.

Key words: Disaster risk management; poverty; social protect; development

Poverty and vulnerability to natural hazards are closely linked and mutually reinforced. Disasters result in hardships and distress, potentially and temporarily forcing certain groups to fall into poverty and even leading to more persistent and chronical poverty. Disasters can result in the loss of lives, houses and assets, disrupt normal livelihood, schooling and provision of social services, erode savings and cause health problems and sometimes cause long-term consequences. Disasters can also disrupt ongoing poverty reduction activities and force a diversion of related financial resources into relief, recovery and rehabilitation efforts. Poverty can be further reinforced by deliberate

---

① Sanny R. Jegillos is Regional Programme Coordinator, UNDP Asia-Pacific Regional Center.

risk-averting and livelihood choices that poor households may make.

Poor and socially disadvantaged groups are among the most hazard vulnerable people, particularly for the rural poor and informal urban dwellers, due to their social, cultural, economic and political environment for instance , the sub-standard quality and often dangerous location of housing, lower levels of access to basic services, and uncertain ownership rights, reduced their incentives to manage resources sustainably or invest in structural mitigation measures; more vulnerable livelihoods and limited access to financial resources constrained their ability to diversify livelihoods and recover from post-disaster. The poor can also exacerbate their own risk since limited livelihood opportunities force over-exploitation of the local environment.

For consideration of delegates of this symposium, the following propositions require attention.

1. Improve Understanding of Disaster and Poverty Linkages

1.1 Intensive risk

a. Concentrated in seismically active regions, coastal zones, flood plains and cyclone track zones.

b. Changes over time with changes in vulnerable populations, economic assets and lifeline infrastructure exposure.

c. Causes large-scale catastrophies thus receive formal support in relief and recovery and implement structural mitigation measures.

· In El Salvador, the two earthquakes in 2001 led to an estimated 2.6-3.6 per cent increase in poverty.

· In Honduras, the percentage of poor households increased from 63.1 per cent in March 1998 to 65.9 percent in March 1999 as a consequence of Hurricane Mitch in October 1998. The number of rural households living in extreme poverty or indigence rose by 5.5 percentage points.

· In Vietnam, it is estimated that a further 4-5 percent of the population could be pushed into poverty in the event of a disaster.

· In Aceh, Indonesia, the 2004 tsunami is estimated to have

increased the proportion of people living below poverty line from 30 percent to 50 percent.

1.2 Extensive risk

a. Increasing in Asia due to greater frequency and intensity of extreme climate events.

b. More frequent, dynamic and widespread, affecting livelihoods and poverty.

c. Invisible in official reporting, risks that are unaccounted for and disguise an increasing burden of risk to low-income households and communities.

d. Invisible to official response systems, increase burden of coping with consequences of impact since lack of formal support for response recovery and institutional disaster risk management programmes if not absent.

UNDP APRC (Asia-pacific Regional Center) helped establish loss data base (30 years) and contributed to GAR 2009/2011 Report. The report shows the significance of the losses due to extensive risk as follows:

• 10% more deaths, 80% more affected, 20% more economic loss, 50% more houses damaged, 83% more injured, 45% more schools damaged and 55% more health facilities damaged

• 93% are hydro meteorological hazards (1989-2009) in 20 countries studied

• Comparative analysis at sub-national level reveals that risk is increasing most rapidly in small and medium-sized urban centers with weaker capacities to manage urban growth; deforestation and destruction of coastal ecosystems are magnifying risk; landslide and flood risk at the local level are closely associated with poverty (poor people living in marginal areas with limited resources). For example, in Indonesia and Sri Lanka, mortality risks from landslides are higher in areas with low level of economic development and higher level of poverty.

1.3 Recommendations for improving analysis

a. Build and maintain quality disaster loss databases (including historical).

b. Improve poverty datasets.

c. Disaster risk reduction must include considerations for the poor and vice-versa.

d. Use intensive/extensive analysis to draw attention of policy makers to extensive risks which provide "in time" information for risk accumulation that will eventually result into intensive risk or catastrophic event.

1.4 Example of best practice supported by UNDP APRC and Country Office

The Indonesian Disaster Data and Information Management (DIBI) is based on official governmental data from the year of 1815 to 2009. DIBI is already used as the basis for national policy-making, planning and budgeting in disaster risk reduction and making development planning decisions. For example, the BNPB has already used DIBI to identify hazard-prone areas across Indonesia, in order to prioritize the creation of district-level disaster reduction systems. Within the BAPPENAS, the Directorate for Poverty Eradication has used DIBI to establish priorities within its own and donor-funded programmes. Further work is now ongoing to enhance DIBI to incorporate additional attributes such as school age, children, health status, infrastructure, public facilities, income levels, types of livelihood and spatial planning data.

2. Integration of Social Protection and Disaster Risk Management

The process of integration maybe examined from two perspectives: social protection and disaster risk reduction and we suggest that the following challenges and opportunities for integration should be considered in China.

2.1 Social protection lens

In the region, there are very limited examples of risk or crisis sensitive approach to social protection. Social protection practiced today does not reduce risk, but it is not difficult to see opportunities where instruments can be adapted to enhance individuals' and households' disaster/climate risk resilience along with its primary goals of poverty reduction and human capital development. Many of these social protection programmes have already been delivered on a large scale, but they are not necessarily targeted to disaster-prone communities and in response to specific crisis events. Thus, through adaptations of targeting criteria and timeframes, they can be used to reach out to very large numbers of disaster-prone households and communities.

2.2 Disaster risk reduction and recovery lens

At the moment, disaster risk management activities are limited in scale and are independent activities with little programmatic linkage with social protection programmes. The scope is typically very narrow and is limited to strengthening early warning systems and disaster response capabilities. This is assumed to be the same as climate change adaptation programmes. Post disaster relief and recovery are temporary in nature and are focused on areas affected by natural disasters. Although international relief measures are sometimes directed towards poor and vulnerable people, the majority of public investments in risk reduction, recovery and reconstruction are not aimed at helping the disadvantaged. Institutions involved in these (public works, water resources, agriculture, scientific agencies, civil defence etc.) are not sensitive to poverty reduction goal. They are not automatically helping impoverished population.

2.3 Questions to ponder in this symposium

2.3.1 Social protection

Should the primary purpose of protecting the poor against risk and shocks and ensuring basic living standards in poor regions be recognized as the priority?

2.3.2 Disaster risk management and recovery

Should deliberate sensitivity to the disadvantaged be engineered in managing disaster risk management and recovery programmes?

Many countries may not have developed a comprehensive system of social protection strategies and policies, but have separated development plans and strategies in various sectors including health, education, women and children, agriculture and etc. Hence, it is necessary to review national policy priorities, poverty profile, legal framework, existing programmes and intervention related to social protection, identify gaps, and assess resource availability to determine the approaches to social protection and primary aims that are suitable for the specific needs of the country.

2.4 Proposed Institutional Actions

Integration can be done through the following:

2.4.1 Targeting

This can be done through a combination of targeting criteria i.e. focus on specific groups who live in poverty-stricken regions where are known as areas at risk to natural disasters and focus on time, like by season—in exceptionally dry seasons to cover droughts or during heavy rains to cover floods.

To set up safety nets aiming at meeting basic food needs and/or minima cash income can allow the chronical poor to divert efforts from survival type subsistence strategies (such as distress sale of remaining productive assets) to development type subsistence strategies that include purchase assets and increase savings. These safety nets should be established ahead of time, carefully targeted towards the poor and designed to support rapid recovery, and if possible, to enhance resilience to future hazard events.

2.4.2 Mainstreaming

As poverty is a complex problem that requires multi-dimensional, multi-scale and multi-sectoral responses, disaster risk reduction ought to be mainstreamed rather than remain in its current independent posture. Key actions required to be incorporated into social protection responses can include: an assessment of vulnerability of the programme and its intended beneficiaries to natural disasters; rational and valid decisions and actions should be taken on how to address these risks and post disaster support (relief, recovery) should be planned ahead of time to provide timely assistance, rapid recovery and enhance resilience to future events.

3. Mainstreaming Disaster Risk Reduction into Development that Benefits the Poor

Recommendations from Asia Case Study supported by UNDP APRC: Possible interventions of combining disaster coping and development will benefit the poor:

Strengthening national and state-level social protection programmes to provide lifeline support to the poor while recovering from shocks, and to prevent them from falling into or deepening their poverty.

Promoting economic and livelihood diversification, especially outside the primary sector and other especially vulnerable economic sectors.

Poverty reduction programmes and DRR should not only target at the poor, but also those who could fall into poverty as a result of shocks.

Strengthening institutional, market and credit linkages to enable the development of truly sustainable livelihoods for the poor.

Strengthening local development and the capacities of DRR and community-based disaster management to enhance resilience and address extensive climate-related risks.

Analyzing the costs and benefits of investment in joint or independent poverty reduction and DRR interventions providing economic data for corresponding decision-making.

Promoting and supporting corresponding studies including the impact of disasters and shocks, and their linkage with human development and development interventions in such areas as poverty reduction, housing and infrastructure.

Based on this, the joint-up of national and state-level development and DRR policies at national and state level, and enabling convergence at the levels at which the mitigation actions or risk crystallizes. This would not only lead to reforms of existing programmes but their decentralization, if extensive risk mitigation is the main objective.

Establishing and strengthening capacities of national institutions to monitor the occurrence and impact of disasters, and temporal and spatial trends of vulnerability, risk and poverty at sub-national and local levels.

Monitoring and reporting on the patterns of disaster risk that is related to changing development trends and their induced vulnerabilities.

Integrating disaster risk analysis and mitigation planning into the design and financing of development programmes and infrastructure projects.

Upgrading both rural and urban housing stock, and improving the quality of new buildings and lifeline infrastructure.

Responding systemically to the significant public health challenges that are linked to risk in Asia.

Establishing appropriate disaster vulnerability, risk and poverty reduction monitoring mechanisms and capacities at all levels.

Building a consciousness and capacity to address both intensive and extensive risks in urban areas.

Basis and Prospect of Pilot Poor Villages' Sustainable Livelihoods in Post-reconstruction Period

—An Analysis Based on the Second Annual Comprehensive Assessment Data

Cai Zhihai*

Abstract: Based on the Second Annual Comprehensive Assessment Data of "the Early Recovery and Disaster Risk Management Programme" sponsored by UNDP (United Nations Development and Planning), this article analyzed the basis of sustainable livelihoods of the pilot-reconstruction poor villages in Wenchuan earthquake-stricken areas, and explored the prospect of sustainable livelihoods under the current policies and measures.

Key words: Post-reconstruction period; Pilot poor village; Sustainable livelihoods

Three years has passed since Wenchuan Earthquake happened. According to the decision of the Central Government that "three-year reconstruction missions should be basically completed in two years", the two-year task of post-disaster reconstruction had been finished by the end of September, 2010. Through two years' reconstruction, the

---

* Cai Zhihai, Associate Professor of Center for Poverty and Rural Governance Studies in Central China Normal University.

disaster-stricken areas such as infrastructure, industrial development, democratic governance and self-development capacity have recovered to pre-disaster level. However, the long-term reconstruction and sustainable livelihoods①problem hasn't been solved in the two-year reconstruction. During the reconstruction period, livelihood development should be emphasized in earthquake-stricken areas to realize the combination of post-disaster reconstruction and poverty alleviation and development. The State Council Poverty Leading Group Office of Poverty Alleviation and Development selected 100 pilot villages to explore the basic experience of post-disaster reconstruction in three stages. This paper is based on the July 2010's survey data②which derives from a comprehensive assessment carried out by UNDP's project "the Early Recovery and Disaster Risk Management Programme". It analyzed the basis of sustainable livelihoods of the pilot-reconstruction poor villages in Wenchuan earthquake-stricken areas, explored the prospect of sustainable livelihoods under the current policies and measures.

1. Analysis Based on the Reconstruction Content of Villages

1.1 The indicators indirectly related to livelihoods in pilot poor

---

① Sustainable livelihoods is a concept that is proposed and developed by foreign scholars since the 1980s. Chambers (R. Chambers) and Conway (G. Conway) said in 1992: "A livelihood is sustainable when it can cope with and recover from stresses and shocks and maintain or enhance its capabilities and assets both now and in the future, while not undermining the natural resource base." After this concept was put forward, many scholars and international organizations developed the analytical frameworks of sustainable livelihoods, among which one of the very well-known is the United Kingdom Department for International Development's (DFID) Sustainable Livelihoods Framework.

② The survey started on July 19-26, 2010 in Sichuan, Gansu and Shannxi provinces, selecting eight sample villages from the first batch of 19 pilot reconstruction poor villages that are determined by the State Council Poverty Leading Group Office of Poverty Alleviation and Development to carry out a comprehensive assessment of post-disaster reconstruction. There were a total of 1200 sample households, valid questionnaires were 1143 by screening, and the effective rate is 95.3%. Meanwhile, a lot of qualitative data were collected through field interviews. As one of the project leaders, the author had a full participation in the survey.

villages

Although the livelihoods systems of farmers are affected by many factors, some factors don't directly affect the livelihoods strategies and results, instead they play an indirect role. These factors include infrastructure such as the village roads, drinking water, renewable energy, household power supply, public service facilities (including education, health, entertainment, and consumption) and village environment, etc. According to the assessment data, from the perspective of the farmer's subjective evaluation, we can understand the current reconstruction and grasp the indirect basis of reconstruction of the pilot poor village's sustainable livelihoods.

Table 1   The reconstruction satisfaction rate of indicators indirectly related to livelihoods(%)

	Infrastructure				Public service				Village environment
	Road	Drinking water	Renewable energy	Home power supply	Primary school	Health room	Activity room	Grocery store	Five changes and three construction
Very satisfied	26.0	16.6	18.4	26.3	26.5	9.6	10.9	13.7	17.4
Somewhat satisfied	48.3	54.7	51.7	65.0	57.6	54.0	53.3	59.9	52.8
Medium	8.9	11.2	20.9	5.9	13.0	29.3	28.4	22.0	18.7
Somewhat dissatisfied	11.4	14.1	7.0	2.3	2.5	6.2	6.5	1.3	8.8
Very dissatisfied	5.3	3.4	1.9	0.6	0.4	0.8	0.9	3.1	2.3
Satisfaction score	3.78	3.67	3.77	4.14	4.07	3.65	3.67	3.80	3.74

The evaluation data of Table 1 from the questionnaire survey in July 2010 about the reconstruction of the pilot poor villages show that:

The overall satisfaction rate about the four indicators in infrastructure which are indirectly related to livelihood is over 70%, and the highest satisfaction score① is reconstruction of household power supply facilities, then the construction of roads, water facilities and renewable energy. In these four indicators, the construction of village roads is more significant for livelihoods activities.

As for public service, most of the village schools have a new appearance after post-disaster reconstruction, and these schools become the strongest and the most beautiful buildings in rural areas, so the satisfaction rate about schools is high. However, the influence of basic education on livelihood cannot be achieved in the short term. In contrast, the satisfaction about the reconstruction of village clinic and activity rooms for villagers is slightly lower, but it is undeniable that the physical and mental health of villagers has a great impact on the families and economic conditions.

Villages' environment is primarily reflected in the "Five Transformations and Three Construction" project, that is, transform water supply facilities, roads, kitchens, lavatories, and poultry house facilities; building gardens, pools and houses②. The data indicates that more than 70% of the villagers expressed their satisfaction to this reconstruction projects.

In general, the indicators indirectly related to livelihoods show that the two-year's reconstruction has finished the reconstruction of the road, water and electrical facilities, improved the living environment of villages, upgraded basic education and basic medical care service, and eased the previous tense relationship between human and environment. Improvement in basic living conditions facilitated the production and operation activities for farmer and laid the foundation for the development of sustainable livelihoods in poor villages.

---

① Satisfaction score is calculated by giving "very satisfied" to "very dissatisfied" accordingly "5-1" points, and then weighing them to obtain the average.

② Some parts of "Five Changes and Three Construction" project are the content of reconstruction of infrastructure; however, some parts are directly related to family production and living. Such as building gardens, in fact, it is engaged in courtyard economy, it is a means of livelihood activities.

1.2 The indicators directly related to livelihoods in pilot poor villages

Combination of post-disaster reconstruction and poverty alleviation and development is a clear requirement of Central Party Committee and the State Council toward Wenchuan Earthquake disaster reconstruction. Part of the content of the pilot poor villages' reconstruction directly involves the village livelihoods development, farmers' livelihoods capacity and livelihoods activities. It mainly includes irrigation facilities, basic farmland restoration, the establishment of village-level mutual funds which provide start-up funds for poor famers, and the training of agricultural technologies and the training of labor abilities. According to the assessment, we can judge the foundation of reconstruction of sustainable livelihoods in pilot villages according to the scores of indicators directly related to livelihoods.

Table 2 The reconstruction satisfaction rate of indicators directly related to livelihoods(%)

	Infrastructure		Public service		Capacity construction	
	Irrigation facilities	Basic farmland	Village-level mutual funds	Start-up funds	Agricultural technology training	Labor transfer training
Very satisfied	18.1	12.6	8.2	7.4	7.4	5.5
Somewhat satisfied	43.6	50.4	37.7	28.1	36.6	30.4
Medium	19.3	20.4	35.6	39.4	34.3	39.5
Somewhat dissatisfied	13.1	10.4	14.7	18.5	15.0	15.3
Very dissatisfied	5.9	6.1	3.8	6.6	6.8	9.4
Satisfaction score	3.55	3.53	3.32	3.11	3.23	3.07

The evaluation data of Table 2 were collected from the questionnaire survey in July 2010, from which we can find that:

The satisfaction rate of both irrigation facilities and basic farmland restoration is over 60%. But, if compared with the other four indicators of infrastructure reconstruction mentioned above, the two indicators directly related to production and management are 10% lower in terms of satisfaction rate. The survey finds that some of village irrigation facilities still have not been finished, which undermines the satisfaction rate of villagers to some extent. Of course, irrigation facilities reconstruction is related to the climate and geographical conditions, so not every pilot village has done the work of reconstructing irrigation facilities.

The goal of production development is that it can provide necessary fund for villagers to recover and develop livelihoods, but the survey data shows that villagers are not satisfied with the establishment of village-level mutual funds and provision of start-up production funds, and the overall satisfaction rate is only 45.9% and 35.5%, the satisfaction score is just slightly higher than 3 points. When we started to carry out the assessment, each village had started village-level mutual funds, but it's still running very unsmoothly. The main reason that many villagers did not participate in village-level mutual funds or borrowing money, is that they had no money to buy a share or there wasn't any good development project or the amount of loans is too few. Moreover, both the borrowing rules and the publicity of mutual funds had a negative effect on villager's participation. Even some government officials in poverty alleviation system thought that it's too difficult to operate the mutual fund. They said that many difficulties in operation greatly decreased villagers'enthusiasm to participate in it. The coverage of production start-up funds is also too small, and unfair distribution of resources and other issues can also affect the evaluation of the villagers.

Capacity construction focuses on improving the livelihoods ability of the villagers. Actually, the satisfaction percentage of the villagers is not very high while the proportion of the dissatisfied is more than 20%

despite of the fact that many villages have carried out practical training in agriculture technology and labor transfer. Judging from the field interviews, the reasons for villagers' low satisfaction for capacity construction of livelihoods lies in the following aspects: Firstly, there is no household technology training; secondly, the training is not concrete and becomes a mere form; thirdly, training resource is distributed unfairly.

In general, the indicators' scores of reconstruction directly related to the livelihoods in pilot villages are not high. Although many factors lead to this situation, but to some extent this reflects the combination of post-disaster reconstruction and poverty alleviation and development is not strong, and the direct basis of reconstruction of sustainable livelihoods still needs to be strengthened.

2. Analysis Based on the Current Livelihoods Situation of Villagers

2.1 Villagers' livelihoods sources and expenditure situations

In the assessment, we examined the villagers' annual cash income and main expenditure in 2009, descriptive statistics data of 1143 samples are as follows in Table 3. In Table 3 we can find that in 2009 the average total income of each household is ￥14972.3, and ￥3630.3 per person. The average total expenditure of each household is ￥15157.9, and ￥3675.3 per person. Thus in 2009, household balance is ￥-185.6.

From the perspective of income, wage accounts for 56.9% of the average total household income, the main livelihood strategy for the famers is to migrate to cities and find a job, which is roughly the same as pre-earthquake. And the followings are crop production income, livestock breeding income, receiving cash gifts and government subsidies respectively. The average total income per household is showed in Figure 1. It can be found that by further calculation, in 2009 the total income of those who are identified as reconstruction households (their houses were badly damaged by the earthquake) is higher than the income of maintenance households (their houses suffered damages but can be repaired), the former is ￥15403.6 while the latter is ￥13399.6, the gap between them is ￥2004.0.

Table 3 Villagers' household incomes and expenditures situation in 2009① (￥)

Income	Min	Max	Average value	Standard deviation	Expenditure	Min	Max	Average value	Standard deviation
Crop production	0	60,000	1,676.3	3,101.4	Crop production	0	40,000	793.7	1,510.6
Livestock breeding	0	110,000	817.1	4,009.5	Livestock breeding	0	300,000	1,057.4	11,342.6
Commercial activities	0	100,000	1,751.7	7,513.6	Commercial activities	0	170,000	1,088.1	7,094.5
Wage income	0	80,000	8,320.9	9,576.9	Education fee	0	50,000	2,186.2	4,486.2
Received cash gifts	0	30,000	745.4	2,885.2	Medical fee	0	112,000	2,920.1	7,415.7
Government subsidies	0	48,000	680.5	2,749.1	Daily expenses	0	150,000	5,463.8	6,472.8
Other income	0	80,000	634.7	3,892.2	Bestowed cash gifts	0	112,500	1,671.2	5,633.1
Total income	0	202,000	1,4972.3	15,116.0	Total expenditure	0	302,300	15,157.9	19,028.4

① The statistics is based on 1,143 samples. In income part, in fact, not every family had relative income in every aspect. For example, some families had no wage income because they had not been employed by anyone. Some families were not engaged in commercial activities, so they had no commercial incomes. Therefore, if one aspect is calculated separately, it would be higher than the average income showed in the table, especially in items such as wage incomes, commercial incomes, received cash gifts, because there may be a situation that family villagers who might not work for cash, did not carry out commercial activities, or had no weddings or funerals in 2009. Take working salaries for example, there were 743 households who had at least one family member with wage income in 2009 and the average working income is ￥12,820.5.

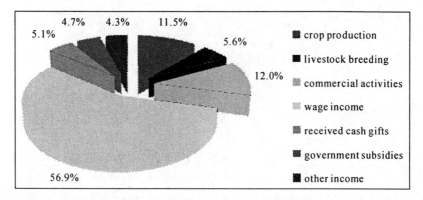

Figure 1   The composition of average total income of a household in 2009

From the view of expenditures①, the main expenditure of villagers was for daily food, clothing and some other essentials, etc., accounting for 36.0% of the average total expenditure, and the followings were medical fee, education fee, bestowed cash gifts, commercial costs, crop production costs and livestock breeding costs respectively (see in Figure 2). Therefore, we could find that production and management fund was very limited for villagers when they had to afford the necessary expenditures such as the daily expenses, medical fee and education fee. By further calculation, reconstruction households' average expenditure was ￥15,229.9 in 2009, while the maintenance households ￥14,895.1, and the gap was ￥334.8, which was very small except for housing expenditures.

Seen from the point of keeping a balance between income and expenditure, there were 538 households that could not make ends meet in the 1,143 samples, accounting for 47.1% of the total samples. On average the expenditure of every household was ￥11,808.1 higher than the income. There were 12 households that made both ends meet, accounting for 1.0%. There were 593 households whose income is higher than expenditure, accounting for 51.9%. On average the income

---

① During assessment, it focuses on investigating villagers, conventional expenditure, while housing expenditure is not a conventional expenditure, so housing expenditure is not considered in 2009 expenditure.

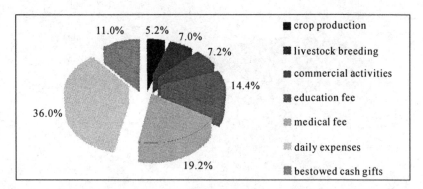

Figure 2 The composition of household average total expenditure in 2009

of these households was ￥10,355.1 higher than the expenditure. The distribution can be seen in Table 4 as follows.

Table 4 Comparison of income and expenditure of households in 2009

	Expenditure＞Income (N=538)		Income＞Expenditure (N=593)	
	Frequency	Percentage	Frequency	Percentage
￥50001 or above	19	3.5	7	1.2
￥20001-50000	54	10.0	69	11.6
￥10001-20000	103	19.1	138	23.3
￥5001-10000	120	22.3	146	24.6
￥1001-5000	191	35.5	174	29.3
￥1000 or below	54	10.0	59	10.0

The investigation about the differences between reconstruction and maintenance households found that there were 438 households which could not make ends meet in the 897 reconstruction households, accounting for 48.8%, and the average household's expenditure is ￥11,127.6 higher than the income. There were 11 households that made both ends meet, accounting for 1.2%. There were 448 households whose income was higher than expenditure, accounting for 49.9%, and the average income was ￥11,226.9 higher than the expenditure. However, there were 100 households which were unable to make ends meet in 246 maintenance households, accounting for 40.7%, and the average expenditure was ￥14,788.6 higher than the income. Only one household managed to make ends meet, accounting for 0.4%. There were 145 households whose income was higher than expenditure, accounting for 58.9%, and the average income is ￥7,661.8 higher than expenditure. The concrete distribution of income and expenditure can be seen in Table 5 as follows.

Table 5　The distribution of income and expenditure for households of reconstruction and maintenance in 2009

	Reconstruction				Maintenance			
	Expenditure＞Income		Income＞Expenditure		Expenditure＞Income		Income＞Expenditure	
	Frequency	Percentage	Frequency	Percentage	Frequency	Percentage	Frequency	Percentage
￥50001 or above	13	3.0	7	1.6	6	6.0	0	0.0
￥20001—50000	44	10.0	59	13.2	7	7.0	10	6.9
￥10001—20000	89	20.3	112	25.0	14	14.0	26	17.9
￥5001—10000	100	22.8	108	24.1	20	20.0	38	26.2
￥1001—5000	147	33.6	119	26.6	44	44.0	55	37.9
￥1000 or below	45	10.3	43	9.6	9	9.0	16	11.0
total	438	100.0	448	100.0	100	100.0	145	100.0

2.2　The condition of household liabilities

A large number of families were unable to make ends meet. Especially, there was a huge housing expense for the reconstruction families. Therefore, many families had a heavy debt burden. Data of evaluation showed that only 241 households completely had no debt in 1,143 households, accounting for 21.1% of the total samples, and the other 902 households, accounting for 78.9%, were in debt to others.

Among the households in debt, 588 households borrowed money from friends and relatives, 765 from formal financial institutions such as credit cooperatives and banks, 22 from mutual-aid funds, 19 from other channels. Different channels of borrowing money can be seen in Table 6. The average household's debt was ￥3,257.7, most of which was borrowed from friends, relatives and banks. By investigating the distribution of the households' debts, it can be found that 77.0% of households having debts ranging from ￥10,001 to 50,000, and the distribution can be seen in Figure 3 as follows.

Table 6 Current household debts situation

	Frequency	Maximum(￥)	Average amount(￥)	Standard deviation
Friends and relatives	588	120,000.0	19,291.9	15,258.7
Credit cooperative/bank	765	300,000.0	22,768.0	16,838.6
Mutual-aid funds	22	23,000.0	6,354.5	6,867.9
Other channels	19	55,000.0	22,157.9	14,477.0
Total	902	400,000.0	32,507.7	23,802.5

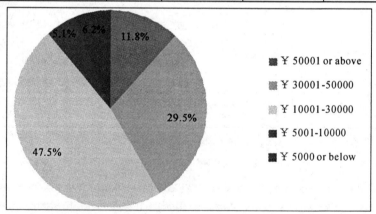

Figure 3 Distribution of average's household's debts

Seen from different types of households, only 108 households had no debt in the 897 reconstruction families, accounting for 12.0%, and other 789 households had debts, accounting for 88.0%. The average household debt was ¥34,008.3. Among them, 700 households borrowed money from credit cooperative or bank, and the average amount was ¥23,171.4; 508 households borrowed money from friends and relatives, and the average amount was ¥19,913.1. 133 out of 246 households had no debts in maintenance households, accounting for 54.1%, and the other 113 households had debts, accounting for 45.9%, and the average household debt was ¥22,030.1. Among them, 65 households' debts came from credit cooperatives or banks, and the average amount was ¥18,423.1; 80 households had an average amount of ¥15,347.5 debts owed to their friends and relatives. The following Table 7 shows the details. Therefore, by comparing them, the debts burden of reconstruction households is ¥12,000.0 higher than that of the maintenance households, and the proportion and average amount of debts borrowed from formal financial institutions were also higher than maintenance households. So the repayment burden of reconstruction households will be very heavy in the next few years, which can weaken their living and production conditions.

Table 7 Reconstruction households' liabilities condition compared with maintenance households

	Reconstruction household (N=897)	Maintenance household (N=246)
Number of households with loan	789 households, 88.0%	113 households, 45.9%
Average loan	¥34,008.3	¥22,030.1
Loan from credit cooperative/bank	700 households, ¥23,171.4	65 households, ¥18,423.1
Loan from friends and relatives	508 households, ¥19,913.1	80 households, ¥15,347.5

2.3 The recovering condition of household livelihoods

Has the framers' current livelihoods recovered to their normal condition or not after two years of post-disaster reconstruction in pilot villages? Are there big changes of their living condition compared with that before the earthquake? These aspects have been investigated in the assessment.

When farmers were asked about "whether their current daily life had recovered to their normal condition or not", 87.7% of those surveyed indicated that they had recovered to normal life, only 12.3% of those surveyed answered "no".

When farmers were asked about "whether their production and management activities recovered to their normal condition or not", 88.1% of those surveyed gave positive answers, and only 11.9% responded negatively. Cross-analysis also shows that the elders were more inclined to think that their production and management activities had recovered, while other factors had no obvious relations with this indicator.

When farmers were asked about "whether their living standard had big changes or not compared with that before the earthquake", 46.4% of those surveyed thought that the living standard had improved, 35.3% of those surveyed held that the living standard had no changes, while 18.3% of those surveyed thought the living standard had descended. Cross-analysis also indicated that some people, including the male, the elders, the head of households, the village cadres, the party members and the maintenance families, were inclined to think that living standard had improved compared with that before the earthquake.

Although farmers' subjective assessment is not bad, some cases in the interview show that the living conditions of certain special groups are still faced with many difficulties. The poverty-stricken pressure resulting from the lack of money is very universal in the poor villages.

2.4 The main difficulties of farmers' livelihoods

We also investigated the current main difficulties of the farmers in production and daily life. The basic condition can be seen in Table 8 as follows. The data showed that most of the farmers thought that they had difficulties in seeking funds, projects, technologies, and information, which are closely related to production. These factors directly affected the farmers' income, and also influenced their living conditions. Meanwhile, deficient living facilities, insufficient food and drinking water are also problems encountered by some farmers. All of these need attention from relevant government departments. At the same time, apart from some internal factors, external environment, especially the change of market, also had directly impacts on the farmers' income in pilot villages.

Table 8   The main difficulties of farmers' production and daily life

	Frequency	Percentage
Lack of production start-up fund	814	71.5
Heavy household debt	799	69.9
No suitable industries	733	64.5
Lack of technological training	720	63.4
Lack of labor transferring information	699	61.9
Deficient production facilities	466	42.4
Incomplete living facilities	362	31.7
Insufficient food	263	23.1
Lack of drinking water	210	18.4

3. Analysis Based on Policies and Measures

3.1 Policies and measures at the national level

The combination of post-disaster reconstruction and poverty alleviation and development is a basic principle for poor villages' post-disaster reconstruction. After two years' reconstruction, great achievements have been made, but it does not mean that rehabilitation and reconstruction of disaster-stricken poor villages have been finished. A key issue in post-reconstruction period is that how to achieve the long-term development of those poor villages so as to truly realize the organic combination of post-disaster reconstruction and poverty alleviation and development.

At the national level, although the specific policy has not been formulated, the State Council explicitly pointed out that "supporting the development of disaster-stricken areas such as Wenchuan" in the Twelfth Five-Year Plan for National Economic and Social Development. And early in the second year after Wenchuan earthquake, the State Council had set out to start projects about livelihoods development in Wenchuan earthquake-stricken poor villages.

> ***Column 1: Wenchuan Earthquake Economic Reconstruction Project***
>
> On May 29, 2010, the State Council and Die Deutsche Gesellschaft für Internationale Zusammenarbeit (GIZ, GTZ is its previous name) held the signing ceremony about Sino-German technical cooperation in Wenchuan earthquake economic reconstruction project.
>
> The key point of Sino-German technical cooperation of post-disaster economic reconstruction project is the capacity construction of livelihoods and sustainable development, thus to explore the mechanism of combining the post-disaster reconstruction with poverty alleviation and development. It aims at using the global network of GIZ, absorbing the effective practice from international society in post-disaster livelihood recovery and economic reconstruction, and summarizing the experience of Wenchuan earthquake-stricken areas to achieve prosperous and sustainable development.
>
> The project content mainly includes three aspects. The first one is post-disaster sustainable reconstruction and income pattern development in poor villages, namely designing and implementing several sustainable reconstruction and income pattern in pilot villages, by which villagers can improve productivity and reduce the vulnerability against future disasters, and demonstrate the significance of "focusing on livelihood in response to natural and man-caused disasters action". The second aspect is capacity construction in poverty alleviation system, namely supporting the capacity construction at national,

provincial, and county levels. The third one is the exchange of development patterns and experience, which will support two earthquake-stricken areas (Wenchuan and Yushu) to carry out experience publicity and international exchange.

3.2 Policies and measures at the provincial level

The local government of Sichuan, Gansu and Shaanxi has fully realized it is a long-term and arduous work to finish the post-disaster reconstruction, and the working focus of post-reconstruction period must shift to the sustainable development of disaster-stricken areas. Guided by highly unified thought, corresponding governmental departments have conducted a lot of research work, and introduced a series of policies and measures aiming at promoting sustainable development in poor villages. In Sichuan Province, as the three-year reconstruction target has been accomplished in two years, Wenchuan earthquake-stricken areas have basically laid the foundation for further development and revitalization. Based on thorough investigation, Sichuan Province began formulating the comprehensive sustainable development program for post-reconstruction period, and prepared for *Wenchuan Earthquake-stricken Areas Development and Revitalization Plan*[①]. And since 2010, Gansu and Shaanxi have also issued related reconstruction policies to help earthquake-stricken areas realize livelihoods restoration, such as increasing infrastructure and public service investments, extending loans repayment period, reducing the loan interest rate, promoting industry support policies and so on. These policies and measures are beneficial for the recovery of livelihoods in poor villages.

---

① According to the latest news on May 15, 2011, the National Development and Reform Commission (NDRC) recently agreed in a formal reply to the *Wenchuan Earthquake-stricken Areas Development and Revitalization Plan*. After the post-disaster rehabilitation and reconstruction were fully completed, the NDRC will continue to support the development of earthquake-stricken areas. This indicates that the *Wenchuan Earthquake-stricken Areas Development and Revitalization Plan* has been a national strategy.

Column 2: Wenchuan Earthquake Earthquake-stricken Areas Development Revitalization Plan (2010 —2020)

On November 9, 2010, in 70th Sichuan Provincial Government Executive Meeting, *Wenchuan Earthquake-stricken Areas Development and Revitalization Plan* was passed. It outlined the path that the development speed of earthquake-stricken areas should be faster than that of the entire province.

The meeting points out that it is of great importance to stick to the overall orientation of speeding up development, scientific development, and sound and fast development, to take the starting point of guaranteeing and improving livelihood of the people, to adhere to the basis of accelerating the industrial revitalization, to enhance the foundation of ecological environmental protection and prevent geological disasters and reduce disaster risk, to increase the support of funding and policies. By all these means, we can consolidate the outcome of post disaster reconstruction, improve the development, speed up the economic revitalization and social development, and enhance the ability of sustainable development in disaster-stricken areas.

The plan puts forward the following goals that, it takes five years to turn the earthquake-stricken areas into a new place where people live in happiness, and both urban and rural are prosperous, human and nature are in harmonious coexistence. Therefore, it will promote economic and social development and help build a solid foundation for sustainable economic and social development and a well-off society in the disaster-stricken areas. The goals of the revitalization plan include: higher development speed than the entire province; better industrial structure than before; full realization of each household's employment; basical eradication of absolute poverty; enhancement of disaster risk reduction capacity.

4. Conclusions and Suggestions

4.1 Main conclusions

From the analysis of the above three aspects, through more than two-year's efforts, the reconstruction of pilot poor villages in Wenchuan earthquake-stricken areas has basically achieved the expectant goals, and the subsequent policies will be favorable to boost the process of sustainable livelihoods in these villages.

As for the villages, the two-year's post-disaster reconstruction has greatly improved the production and living condition of pilot villages, and has provided more beneficial conditions for sustainable livelihoods than before the earthquake, which were mainly embodied in the construction of road, drinking water facilities, electricity, renewable energy infrastructure, public services and improvement of environmental conditions. However, in the reconstruction, famers' satisfaction rate of the reconstruction projects directly related to livelihoods recovery was not high, such as the insufficient reconstruction of irrigation facilities, recovery of basic farmlands, production development and capacity construction. All these imply that the reconstruction outcome should be further strengthened.

As for the households, after reconstruction, nearly half of the rural households were still unable to make ends meet. In particular, reconstruction households were more serious than maintenance households. Nearly 80% of the households had a debt, and the average amount reached ￥32,500. Besides, the debts of reconstruction households were higher than maintenance households. Although most households thought that their production and daily life had recovered, they still had difficulties in seeking funds, projects, technologies, information, and production facilities, which were directly related to family income, and would greatly influence farmers' living conditions.

As for policies and measures, although the livelihoods reconstruction policies at the national level have not been issued, some relevant projects have been conducted. Local governments have issued plans of livelihoods development and economic revitalization, which would exert great impact on the sustainable livelihoods development of poor villages in disaster-stricken areas.

4.2 Several suggestions

(1) The basic principle of the combination of the post-disaster reconstruction and poverty alleviation and development should be further implemented in post-reconstruction period. The governments should fully realize the post-disaster reconstruction is a long-term and arduous task and their work focus should be shifted to the sustainable livelihoods development of farmers.

(2) It is necessary to issue the national level reconstruction policies about livelihoods recovery in disaster-stricken poor villages in the post-reconstruction period, and to conduct unified directions to the long-term livelihoods development in the three affected provinces.

(3) Discuss the cohesion between the sustainable livelihood development paths in poor villages and the regular means of poverty alleviation and development, especially the effective ways of resource integration and mechanism coordination.

(4) Attentions should be focused on the differences between reconstruction households and maintenance households, and give the appropriate preferential policies and financial support for farmers with high debts.

References:

[1] Chambers, R., Conway, G. Sustainable Rural Livelihoods: Practical Concepts for the 21st Century[P]. Brighton: Institute of Development Studies, 1992.

[2] DFID. Sustainable Livelihoods Guidance Sheets[R]. London: DFID, 1999.

[3] Huang Chengwei, Thomas Bonschab. An Evaluation of the Social Impact of the Implementation of the Overall Plan for Wenchuan Post-earthquake Restoration and Reconstruction[M]. Beijing: Social Sciences Academic Press, 2010.

[4] UNDP. Comprehensive Assessment Report of the Project of "Early Recovery and Disaster Risk Management Programme", 2010.

Significance and Benefit Evaluation of the Project of Returning Farmland to Forest and Grassland in Western Poverty-stricken Areas

—Taking the Longnan City in Gansu Province as an Example

Wang Jianbing Tian Qing[*]

Abstract: Implementing the project of Returning Farmland to Forest and Grassland is an important task and effective measure of poverty alleviation in ecologically fragile areas. Based on the investigation of post-disaster reconstruction in Longnan city, this article considers the poverty alleviation and ecological management comprehensively. Firstly, this paper analyzes the significances of implementing the policy of Returning Farmland to Forest and Grassland on the aspects of ecological safety, food security and the post-disaster reconstruction in Longnan. Secondly, it dissects the benefits of ecological restoration, farmer income increasing and adjustment of agricultural structure of production in that area, which will provide a theoretical reference for impoverished areas to get rid of poverty to become rich and gain ecological rehabilitation.

Key words: Poor Areas; Returning Farmland to Forest and

---

[*] Doctor Wang Jianbing is the Vice-director of the Institute for Rural Development and an associate research fellow of Centre for Research on Poverty Issues in Gansu Provincial Academy of Social Sciences; Tian Qing works in Gansu Provincial Academy of Social Sciences.

Grassland; Significance and Benefit

The project of Returning Farmland to Forest and Grassland (RFFG) aims at replacing agricultural production in sloping farmland that causes serious soil erosion with forestry and grassland production for vegetation recovery, which can be conducive to improving ecological environment of western region, adjusting the agricultural structure and promoting sustainable development of society. The project of RFFG is an opportunity for western region to adjust regional economic structure. Through utilizing national policies effectively, the western region reconstructs the economic structure, especially agricultural structure[1]. The project of RFFG is implemented as a project of emergency rescue. It's a proper timing for the government to carry out the new agricultural policies which will ease the ecological pressure and increase the profit[2]. In 1999, the government launched a pilot program in Sichuan, Gansu and Shannxi. In 2000, the State Council approved a pilot plan in 174 counties in 13 provinces. The project has achieved gradual victory after being implemented for more than 10 years. The policy of RFFG is of great significance in improving environment, optimizing rural industrial structure, promoting the development of western regions and comprehensive construction of a well-off society[3].

1. Basic Condition

Longnan city is located in Qin-ba mountain area, and it is the only part of Gansu Province in Yangtze River basin. The climate there is in transitional zone from sub-tropics to warm temperate, and it is of

---

[1] Li Shuangjiang, et al. *Discussion on the Mechanism of Transferring Plough to Forest (or Grass)*. Journal of Arid Land Resources & Environment, 2004(5).

[2] Xu Jintao, et al. *The Sustainability of Converting the Land for Forestry and Pasture*. International Economic Review, 2002(2).

[3] Cao Xinjing, et al. *Discussion on Grass and Stock Industry Development after Reforestation in Southern Mountain in Ningxia*. Journal of Arid Land Resources & Environment, 2007(7).

vertical distribution, so the regional difference is obvious. The area cultivatable land per person is small, and the distribution of rainfall is not well balanced too, so droughts and floods frequently happen. Longnan is one of the four major regions where landslides and debris flows frequently occur. The frequency of occurrence and damage level of geologic disasters rank in forefront nationwide. Longnan is the severe damaged region by Wenchuan earthquake, with 195 towns and counties, 2,343 villages, 425,800 households, 1.7476 million people being affected by the earthquake. The direct economic loss came to 42.261 billion yuan, and the amount of economic loss in rural areas reached 20.66 billion yuan accounting for half of total direct economic loss of the whole city. Longnan is also one of the most impoverished regions in Gansu. 7 of the 9 counties in Longnan are state-supported impoverished counties, and 2,106 of the 3,237 villages are impoverished villages. Through kinds of state poverty reduction policies, the poverty population has dropped from 1.2888 million in 2000 to 0.7689 million in 2009. The impoverishment rate dropped from 53.35% in 2000 to 31.21%. The average net income of one farmer has increased from 956 yuan in 2000 to 1995 yuan in 2009.

2. Significance of RFFG

2.1 The strategic requirement of ecological security

Longnan is one of four major regions in China where landslides and debris flows frequently occur. 90% of 5.52 million hectares of farmland are steep sloping land, and 280 thousand hectares of them are inclined more than 25 degrees. These sloping land is characterized by sterile, loose soil, and serious soil erosion. According to related information, the soil erosion per year reaches 160 million tons in Longnan which is 25% of sediment inflow from Jialing River to Yangtze River. Thus, to implement the project of RFFG is not only the requirement of ecological civilization construction, but also the strategic choice of reducing sediment inflow to Yangtze River to safeguard the safety of Three Gorges Project and surrounding environment.

2.2 The realistic choice of food security

Longnan is the main over-summering area of wheat stripe rust in China which is a great threat to local wheat production and it also spreads to surrounding late-ripening winter wheat and spring wheat areas. The complicated geographical and ecological characteristics make it possible for wheat stripe rust to survive summer and winter in this area, thus completing the annual cycle in small areas. After the new pathotype regenerates, Longnan becomes the main spreading source to the whole country because of its special geographic condition. In Longnan large areas of wheat are planted above the altitude of 1,600m, the wheat stripe rust can survive the hottest summer with an average temperature no more than 22℃ in those areas in volunteer seedlings or late-maturing winter wheat, making Longnan become the most important over-summer area[①]. Since there is no effective treatment for wheat stripe rust, it is pandemic easily in large areas. Therefore, to return large area of farmland to forest is the only way to prevent the wheat stripe rust, optimize agricultural structure, which is meaningful to the economic development of Longnan and the safety of national wheat production.

2.3 The means of consolidating the achievement of post-disaster reconstruction

Longnan is one of the severely damaged regions in Wenchuan earthquake, in which 7 of 9 counties were severely damaged. The earthquake caused wide-spread debris flows and landslides which made the sloping land totally unsuitable for production and cultivating. Meanwhile, because 243 villages must be moved to Sichuan to be reconstructed, the existing homestead, sloping land around villages has been abandoned. According to incomplete statistics, 107,000 homesteads and sloping land can't be cultivated again due to the earthquake. However, although the farmland has lost the value of

---

① Zhou Xiangchun. Gene *Distribution of Control of Stripe-rust on Wheat in Longnan Region*. Gansu Farming, 2007(6).

cultivating, it is the best area to be returned to forest and the main area for reconstruction to develop a pillar industry and increase income.

2.4 The best way to adjust rural industry structure

"Eighty percent of Longnan are mountains, ten percent water, and another ten percent farmland", is a typical and vivid portrayal of the landscape of Longnan. The mountains is large and grooves are deep, the soil is so poor that the yield of sloping land is less than 150kg per mu. However, the advantage of climate condition (combining subtropical and warm temperate climate, and abundant rainfall in Longnan make it fit for planting more than 300 kinds of cash crops, such as walnuts, Zanthoxylum, olive, tea, ginkgo, tung, kiwifruit, etc. The mountainous area is large enough for afforestation with optimistic prospect. In the program of poverty reduction, the strategy of Returning Farmland to Forest and Grassland should be continued which will contribute to improving environment, optimizing agricultural and industrial structure and increasing farmer's income in the earthquake-stricken area.

3. Benefits of RFFG

3.1 Improving ecological condition.

Through implementing the project of RFFG, the increasing forest area has reached 215,000 square kms, and the forest coverage has increased by 1.5 percent. Especially, by returning the sloping land, the forest soil erosion has been restrained efficiently. The area of soil erosion in Longnan has decreased to a large degree compared with that of 1999. According to the research of relevant departments, after returning the sloping land with 400mm rainfall to forest, the soil erosion has decreased 45-75 tons per hectare and 225 cubic metres of water have been impounded. The amount of soil erosion in Longnan has reduced 5 million tons and impounded more than 15 million cubic metres water. The implement of project has accomplished the goal that water is stored in the mountain and soil in the grooves, and the ecological condition in the region has improved.

3.2 Increasing farmer's income efficiently

The government provided subsidies for the farmers in the process of implementing the project of Returning Farmland to Forest. According to the standard of Yangtze River basin, every farmer got about 150kg of food (210 yuan) and 20 yuan of medical subsidy. During the ten years of project implementing, 268,000 households, 1.129 million farmers had received 13.73 billion yuan in cash, 5,123 yuan per household. The farmers involved in the project had not only solved the need of food, but also liberated the labor force to do diversified business, sideline production or move out for work, which broaden the ways of increasing income. According to the statistics, the project of Returning Farmland to Forest made 250,000 farmars become surplus labor, 15,000 of whom went to urban areas for work, and the total income reached 6 billion yuan (4,000 yuan per person).

3.3 Adjusting the rural industry structure

Due to the implementation of this project, some ecologically important slope land with low and non-stabilized food production has been under control. The utilization rate of land has increased, and the agricultural structure has been successfully adjusted. Accordingly, the mode of agricultural production has been transferred back to the traditional one. During the past ten years the project has developed 74,000 hectare economic forest bases, which makes up 33% of 223,000 hectares of the project's total area, and resulted in a large number of industrially characteristic towns and villages. Especially after Chinese prickly ash, walnut, olive oil, tea and other featured economic forest were brought into the policy of RFFG subsidy range, the mass had high enthusiasm for returning farmland to forest, which ensured the effects of the RFFG project. For example, Wudu district, by adjusting industrial structure, the income from economic forest increase from 32 million yuan in 1999 to 280.18 million yuan in 2009 which accounts for 38.4% of agricultural gross output, the per farmer's economic forest income reached 256 yuan, which is 41% of net income per capita. Through implementing the project of RFFG, the extensive cultivation is turned into modern cultivation which is of intensive cultivation with variety, high quality and efficiency. The project also promotes the development of the tertiary industry such as rural processing industry, transportation industry, service industry and eco-tourism, and it also facilitates to solve issues concerning agriculture, countryside and farmers.

3.4 Accelerating the development speed of non-public forestry industry

The subsidy policy of RFFG activates the public's devotion to ecological construction. Longnan has developed 260 forestry households; the total forest area has reached 27,000 hectare, which accelerated the step of ecological construction in Longnan. Liangdang County has developed 15 non-public forestry households that own more than 1,000 mu forests, and the government has auctioned and contracted 7,000 hectare barren hills for planting forest, and 20 million yuan has been invested. The other counties have developed various forms of non-public forestry industry such as eco-tourism and ecological farm which promoted the development of non-public forestry industry.

3.5 Building up public ecological consciousness

Enhancing the protection and construction of ecology has been the common sense among communities. The cadres and common people have deep awareness that the project of RFFG is not only the effective way to improve environment or construct beautiful hometown, but also the means of becoming well-off and solving issues concerning agriculture, countryside and farmers. According to the special survey of Chinese Academy of Sciences, more than 90% of farmers support the project of RFFG. Through the implementation of the project and relevant policies, the masses in Longnan aware that it is of great strategic significance for the government to subsidize the ecological construction. They participate actively in the ecological construction and the development of characteristic forestry industry. The project of RFFG has been the most positive force in promoting ecology construction and development of rural economy.

4. Conclusion

The practice has proved that implementing project of RFFG is an important task and effective way of poverty alleviation in the ecologically fragile regions. To implement the project is of great significance in upgrading the ecology condition in ecologically fragile regions, adjusting agricultural industrial structure, increasing farmers' income, and decreasing the costs of immigration. In recent years, the masses have benefited from the project, and they demand returning more farmland to forest to change the poor and backward state in rural areas. Especially after the Wenchuan earthquake, the request for continuing the project is urgent.

Suggestion: Firstly, adjust the short-term compensation policy to long-term policy. Secondly, on the basis of investigation and analysis on reinforcing the momentum to implement the project of RFFG, the government should increase the standard of subsidy and enlarge the area of RFFG. Thirdly, seize the opportunity of reform of forest right system, finish the adjustment of the attribute of farmland to attribute of forest, erase the ecological pressure and liberate the productive force. Fourthly, enrich the content of practical technique training, promote employment for development, implement benign ecological migration, broaden the source of farmer's income, and achieve the win-win goal that farmers become rich and the ecologically fragile regions gain self-healing capacity.

(**Translated by Cai Zhihai**)

Japan's Disaster Management and Its Implications for China

Cheng Guangshuai   Liang Hui*

Abstract: As a disaster-prone country, Japan has accumulated rich experience in disaster management. Main experience includes developing a relatively comprehensive legal system for disaster risk reduction, building disaster management agencies conducted by the Prime Minister's Office, formulating comprehensive disaster risk reduction management system for planning and information. China, in the social transformation, not only face frequent natural disasters, but also the occurrence of social emergencies is relatively high. More importantly, China's disaster management system still need to be improved. Thus Japan's experience in disaster management has significant implications for China. This article reviews Japan's disaster management system and its experience, and puts forward localized innovative policy proposals for China's disaster management.

Key words: Disaster Management; Disaster Risk Reduction; Localized Innovation

1. Introduction

Since the beginning of 21st century, there is an upward trend in

---

* Cheng Guangshuai, teacher of College of Public Administration, Zhongnan University of Economics and Law, Doctor of Economics; Liang Hui, teacher of College of Public Administration, Zhongnan University of Economics and Law, Doctor of Economics.

global natural disasters and its death rate tended to rise. In recent years, there has been more diseases such as Indian Ocean tsunami, "SARS" event, Wenchuan earthquake, H1N1 flu. Besides, recent Tohoku Earthquake and other large disasters followed one after another, resulting in a large number of casualties, property damage and serious social disorder. Disaster and its management have become public issues that every government and every citizen have to pay attention to. With the rapid growth of Chinese economic development, the number of natural disasters and losses caused by disaster are rising①(Lu Cong, Wei Yiming, Fan Ying, Xu Weixuan, 2002). In 2004, National Research Center for Science & Technology conducted a social survey in western China, which showed that in more than 2,700 communities involved in the survey, nearly half suffered serious natural disasters② (Zhao Yandong, 2007). For China, development is the top priority. So disaster management is particularly urgent and important. In a sense, disaster and development are two sides of a coin. On the one hand, destruction of environment and resources caused during the development is one of the main reasons for increasingly intensifying disaster; on the other hand, sustainable development also requires disaster risk reduction internally③ (Tong Xing, Zhang Haibo, 2010). As a developing country, China's disaster management system is gradually improving. Therefore, it is a matter of urgency to learn from disaster management experience of other advanced countries and strengthen disaster management system construction, which are of extremely important theoretical and political implications for sustainable development and construction of harmonious society in China.

Japan is located in the marginal areas between the Asian continent

---

① Lu Cong, Wei Yiming, Fan Ying, Xu Weixuan. *Economic Impact of Disasters on the Quantitative Analysis Model and Its Application*. Journal of Natural Disasters, 2002, No. 4.

② Zhao Yandong. *Social Capital and Post-disaster Recovery*. Sociological Studies, 2007, No. 5.

③ Tong Xing, Zhang Haibo. *Disaster Management Framework Based on Analysis of China's Problems*. China Social Sciences, 2010, No. 1.

and the Pacific Ocean, where disasters and volcanic activities occur frequently. In Japan, disasters cause heavy casualties and property losses every year. Therefore, Japanese government and the whole society pay special attention to disasters. Although Tohoku Earthquake occurred in Japan in March this year caused serious damages, Japanese people remained calm and orderly in front of the disaster. Besides, Japanese government responded to the disaster efficiently. All of these had left a deep impression on every Chinese. The reason is that frequent natural disasters not only improved the national sense of crisis in Japan, but also led to a comprehensive set of structural integrity, functional and efficient operation of the earthquake disaster relief mechanisms to ensure relevant parties' timely and effectivly response when the disaster occurred. Thus, learning from Japanese experience gained in the process of fighting against disasters has important referencial value for improving disaster management system of our country.

2. Japan's Disaster Management System

2.1 Evolution of Japan's disaster management system

Japan's disaster management system can be traced back to the law of *Prepare against Natural Disasters Reserve Act* formulated in 1880, the main purpose of which is to ensure enough food and materials supply in the event of disaster or famine. In 1961, Japanese government formulated *Disaster Countermeasures Basic Law* and established comprehensive disaster management system initially. However, since the middle 1990s, successive major disasters such as Kobe Earthquake in 1995 and Russian oil tanker struck a rock and spilled in 1997, not only exposed Japan's deficiencies in original disaster management system, but also contributed to the establishment of Japan's comprehensive national system in crisis management.

In April 1998, reform of the Cabinet Secretariat took place, setting up "Cabinet Crisis Management Supervisor", which the official rank is the same as Deputy Chief Cabinet Secretary, and it governed "Crisis Management and Security Office". The main role of "Cabinet Crisis

Management Supervisor" is to be responsible for assessing damages, coordinating emergency measures that have been issued by central government departments, and take appropriate policies by assisting the Prime Minister and the Chief Cabinet Secretary. Its daily responsibilities are to communicate with experts at home and abroad, study and develop a variety of crisis management measures, check and improve crisis management mechanism. After the reform of central institution in 2001, the crisis management command of the Prime Minister was further enhanced, so was the coordinating power and position of the Cabinet Secretariat. Moreover, all Japan crisis management departments' role in disaster reduction was strengthened, and the Prime Minister took the responsibility of "Central Disaster Management Council" directly as its president. In 2002, Japanese government applied the latest technology and equipment to upgrade the "Crisis Management Command Center" located in Prime Minister's official residence.

Under the new system, Japanese government has upgraded disaster reduction to national crisis management level, directly under the governance of the Prime Minister, and formed a comprehensive system including daily administrative management, crisis management and mass disaster management. Meanwhile, a modernized command system took shape ①(Yuan Yi, 2004; Cabinet Office, 2009), shown in Figure 1.

2.2　Operating process of Japan's disaster management system

Japanese operating system in disaster management can be hierarchically divided into three parts: the Cabinet Office, the Cabinet Secretariat, and each ministry. The Cabinet Office is responsible for planning, for instance, policy-making. The Cabinet Secretariat has real command functions such as crisis management response. For each ministry, its main function is to perform actual implementation, promote cooperation among administrative bodies, public institutions

---

① Yuan Yi. *Japan's Administrative System and Reduction Plans for Disaster Management*. Disaster Reduction in China, 2004, No. 12; Cabinet Office, 災害プレースタッフ研修契約情報, Heisei 21 (2009).

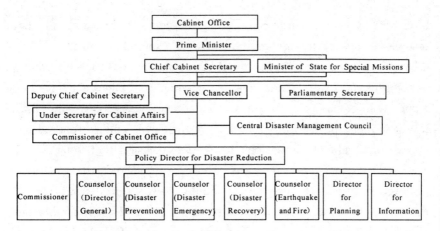

Figure 1 Japanese disaster management command system
(Source: Cabinet Office: 災害プレースタッフ研修契約情報, Heisei 21 (2009)).
and mass groups①(Xiong Guanghua, Wu Xiuguang, Ye Junxing are on duty 2010).

When disaster of emergency occurs, the Cabinet Information Center and the Crisis Management Command Center(These two bodies are on duty 24 hours around the clock, to ensure timely receiving and conveying disaster information) could have first-time access to disaster information via mass media, non-governmental public bodies and relevant government ministries, and then report these issues to the Prime Minister, the Chief Cabinet Secretary, the Deputy Chief Cabinet Secretary and the Cabinet Crisis Management Supervisor respectively. Under normal circumstances, it is not necessary that Prime Minister dispose the problems personally, but the countermeasures office of the Prime Minister's Office will be set up according to different disaster levels. Only when a disaster or emergency related to national security occurs, a conference will be held by the Prime Minister.

Of course, emergency meeting will be different depending on the

---

① Xiong Guanghua, Wu Xiuguang, Ye Junxing et al. *The Evolution of Taiwan's Disaster Prevention and Relief System*; The 2010 Cross-straits Public Governance Forum—Public Administration, Disaster Response and Crisis Management. Social Science Literature Publishing House, 2011.

level of the event. First of all, if emergency information is about terrorist attack or foreign military attack, "special committee of state affaires" and "security council" will be set up. Then, if the information is about crisis of food and health, meeting of "emergency response team" will be needed. When the emergency level rises, the Prime Minister must be informed, and then the issue will be upgraded to "provisional cabinet discussion" level. The decision that whether "the Special Disaster Countermeasures Headquarters" or "the Emergency Countermeasures Headquarters" should be set up depends on the disaster. Thus, the Cabinet crisis management supervisor of this early emergency response system will recommend the Minister for Disaster Reduction or the Chief Cabinet Secretary to serve as head of the Headquarter of Countermeasures. In addition, when emergency need quick and flexible response in the disaster-stricken spots, the "Special Disaster Responding On-the-spot Department" or " Emergency Responding On-the-spot Department" can be set up.

3. Japan's Experience in Disaster Management

3.1 Identify relevant responsibility in disaster management

Early in 1961, Japan developed "Disaster Countermeasures Basic Law", which regulated disaster prevention countermeasures and responsibilities in reconstruction process of administrative agencies after the disaster, and also formulated operating system and planning and financial measures. It specified different groups' disaster prevention responsibilities from the government to common people, promoted comprehensive disaster management in administration and financial assistance, and established special laws, relevant regulations related to preparation and emergency response in disaster period so that post-disaster reconstruction will have legal basis, and various activities can be regulated. All of these promoted the rapid development of disaster management system.

Specifically speaking: (1) Central government's disaster prevention responsibility is to formulate and implement national disaster plan and countermeasures. (2) Japan practices a local autonomous system, regional disaster prevention must, therefore, rely on local government's financial and material resources. (3) Responsibility of city, street and village is to ensure that residents' life and property should not be destroyed in disasters, obtaining relevant authorities and other common groups' assistance as much as possible, formulating disaster prevention plan that is suitable for the region, and promoting implementation of various types of programs stated in plan. (4) Responsibilities of government departments at all levels are to formulate disaster prevention plan related to local department or business and implement disaster reduction activities under the regulation of the *Disaster Countermeasures Basic Law* and other laws related to disaster reduction. At the same time, in order to ensure central and local government implement disaster prevention plan successfully, the designated public authorities have the responsibility and obligation to provide the local government with assistance. (5)Citizen's responsibility is also defined in the *Anti-disaster Basic Law* in which defines the responsibilities of common groups in the local public bodies, managers of disaster prevention facilities and the common citizens, as well as formulating reward and punishment provisions for those officers and citizens according to their performances in the process of disaster prevention and mitigation.

3.2 Perfect organization system in disaster management

Japan also built various types of specialized agencies in disaster management while governments at all levels are building relevant disaster responding systems. In order to improve the efficiency of disaster prevention and mitigation, the Japanese government established a new cabinet organization in 2001, in which the former Ministry of Land, which is in charge of disaster management, merged with the Ministry of Transportation, the Ministry of Construction and the Hokkaido Development Bureau into the "Ministry of Land, Infrastructure, Transportation and Tourism". At the same time "crisis management · disaster management minister" post was set in the new

Cabinet for overall disaster rescue and reduction operations among different agencies. In addition, the Japanese government upgraded the level of the Disaster Prevention Bureau that is originally under the control of the Ministry of Land and set up a full-time post named "Policy Director for Disaster Reduction". The new Cabinet's responsibility for disaster reduction is mainly served by five counselors including: director-general for disaster reduction, counselor for disaster prevention, counselor for disaster responding, counselor for disaster restoration and counselor for disaster contingency measures.

Japan's disaster management system is divided into the central level, the prefectural level and the municipal level. Usually it holds disaster prevention conferences, and will set up Disaster and Countermeasures Headquarters when disaster occurs. If terrible disaster or abnormal disaster happens, the "Special Disaster Countermeasures Headquarters" and "Emergency Disaster Countermeasures Headquarters" will temporarily be set up by the Prime Minister in the Prime Minister Office. Meanwhile, "Disaster Responding On-the-spot Department" can be set up in order to make it convenient to deal with issues on disaster spots.

3.3 Develop comprehensive disaster prevention plan

Japan's "Disaster Countermeasures Basic Law" made specific regulations for the responsibility of the governments at all levels to formulate disaster plans, in which the central government's role is to develop basic national disaster prevention plans, relevant governmental departments are to be responsible for developing disaster prevention plans related to their business; the local government's role is to develop local disaster prevention plans in its administrative area. Basic disaster prevention plan is made by the Central Disaster Management Council, clearly setting up disaster prevention organizational systems and basic guidelines to promote disaster prevention programs, conduct disaster recovery and reconstruction timely. Besides, they also formulate and specify mitigation research plans, preferential terms of disaster management business and local plans. Disaster prevention business plan is formulated by certain government agencies and public institutions to ensure that agencies should take mitigation countermeasures. Local disaster prevention plan is formulated in the charge of prefectures and cities, villages and towns' meeting to ensure mitigation measures be taken by local disaster prevention organizations. Generally speaking, according to the types of disaster every region formulated specific disaster prevention plans including earthquakes, snowstorms, fires, dangerous material incidents, accidental emergencies. Its outstanding feature is to forecast specific natural disaster that may happen in the area, including staff victimization, building damage, fire and the number of people in need of asylum, etc., to make sure that all have done relatively specific estimation[1]. (Yuan Yi, 2004). In this case, some discharge loss can be placed to formulate disaster prevention counter measures for all types of places.

3.4 Building unimpeded information network in disaster management

First of all, Japan established information disaster management under the Cabinet Office, with Cabinet Intelligence Research Office as the central part, to take charge of gathering intelligences, compilation, analysis and utilization, etc., and therefore enforced government's centralized control toward disaster information. Secondly, set up interwined information organization from central to local administration in order to assist authorized information management organizations to make collection, compilation, analysis, and comprehensive utilization in disaster information. At the central level, set up intelligence headquarter of information management organization in relevant agencies as the Ministry of Land, Infrastructure, Transportation and Tourism. In addition to central ministries and agencies that have disaster intelligence transmission systems, relevant agencies in relevant organizations and local government bodies have also established their

---

[1] YuanYi. *Japan's Legal System of Disaster Management*. Disaster Reduction in China, 2004, No. 12.

own intelligence transmission systems for disaster prevention. Finally, according to the characteristics of new technologies involved in digital information, networking and other high-teches, the use of satellites, fixed camera, long-distance image transfer device and small spacecraft (UAV) technology, Japanese government applied well developed communication system to disaster management, and established famous Phoenix disaster management system ①(Li Jun, Nie Yingde, 2009).

4. Japan's Disaster Management Implications and China's Localized Innovations

Japan is a developed country which has accomplished social transformation, and its social conflicts are relatively not very prominent. But the situation in China is different. Although natural disaster itself brings destructive consequence. More importantly, China is under the social transformation, natural disasters will not only cause life and property losses, but the greater risk is that massive social conflicts will arise if poor relief work isn't handled properly. Meanwhile, China's social transformation has accumulated a large number of social problems that put forward higher requirements for China's disaster management. However, as for disaster management, China's disaster emergency response mechanisms and response capabilities remain to be improved, and disaster management system is not smooth② (Li Xueju, 2004). So according to China's national conditions, we must take localized innovations to disaster management as well as drawing Japan's disaster management experience.

4.1 Change disaster management concepts and ideas

With the increase of disasters and its destructive power, people's disaster consciousness is gradually deepened and two common senses

---

① Li Jun, NieYingde. *Japan's Disaster Information System and Its Operation: Experience and Implications.* Around Southeast Asia, 2009, No. 2.

② Li Xueju. *China's Natural Disaster and Disaster Management.* Disaster Reduction in China, 2004, No. 6.

formed. On the one hand, natural disasters' destructive power depends on the interaction between nature and the society, so humans should not only concern the causes of disasters and engineering defense, but more concern about what can be done in response to disasters. On the other hand, social characters of natural disasters is beyond its natural characters and begins to take the dominate position. The main view held by this essay is the content of large-scale disasters. This is basically the same with the definition defined by our *Law of the People's Republic of China on Emergency Response* that disaster is what suddenly happened, did or might do serious harmness to the society, including natural disasters, accidents, disasters, public health and social safety incidents, thus we need to take measures to respond to them. In addition, many studies have found that one of the serious consequences of disaster is that it increased the possibility of poverty, and affected disadvantaged groups particularly. At the same time, the China's large-scale disasters mainly occur in ecologically fragile and poor areas in western region, thus government's disaster management must combine disaster prevention and mitigation with poverty reduction, not only concerning disaster recovery and reconstruction work in short term, but also paying attention to the long-term sustainable development of disaster area and affected groups.

On this basis, China's disaster management should establish several concepts and ideas. Firstly, it should be an overall management, which not only concerns natural properties, but also take into account social attributes. Secondly, it should be a whole-process management, not only manage the environment caused by disasters, but also manage the social consequences caused by disaster management. Finally, it should be a comprehensive management, not only be limited to disaster management itself, but be forwardly extended to prevention management and backwardly extended to post-disaster emergency management. Disaster management system built on these concepts correspondingly has the following three characteristics ①(Tong Xing, Zhang Haibo, 2010): (1) Systematic. Disaster management does not only control the situation and minimize losses, but also rebuild government image, enhances government legitimacy, and even takes this opportunity to promote social reform, optimizes management structure to achieve long-term stability of society. (2) Dynamic. Polymorphism disaster management system ensures that according to specific condition different relevant response measures should be taken in case of unpredictable disasters rather than following the old-fashioned relief mechanism. (3) Initiative. Unexpected events can cause casualties and property losses and social disorder, disaster management initiative is embodied in the preventive management and responsive management, minimizing disaster losses.

4.2 Clearly divide business relationship between China National Committee for Disaster Reduction and other disaster management departments such as the Ministry of Civil Affairs

As we can see in the website of China National Committee for Disaster Reduction②, its primary responsibility is to "study and formulate national disaster reduction policies and programs, coordinate and develop major disaster reduction activities, guide local government's disaster reduction, promote international exchanges and cooperation in disaster reduction. Specific work is undertaken by the Ministry of Civil Affairs". This indicates that China National Committee for Disaster Reduction is only a coordinating institution. Considering that China currently is in a disaster-and-risk-prone period, function of China National Committee for Disaster Reduction should redefine, and clarify its relationship with other specific executive departments.

Firstly, according to *Law of the People's Republic of China on Emergency Response* promulgated in 2007, meaning of disasters is no

---

① Tong Xing, Zhang Haibo. *Disaster Management Framework Based on Analysis of China's Problems*. China Social Sciences, 2010, No. 1.
② http://www.jianzai.gov.cn/.

longer limited to natural disasters, public security, food safety and group events are also included in the scope of the conception of disasters. Thus, the functions of China National Committee for Disaster Reduction should not be limited to responding to natural disasters, but also should regard mass disturbances, accidents, public security as its duties and integrate various types of resources. Moreover, in China, the Ministry of Civil Affairs, the State Administration of Work Safety, the State Food and Drug Administration, Ministry of Public Security, armed police and military all have separate specific duties in disaster responding, some of which overlap. This is not conducive to efficient allocation of national disaster management resources and may undermine the possible positive effect. Therefore, we should reconsider the relationship between disaster management departments of the government and China National Committee for Disaster Reduction, transforming China National Committee for Disaster Reduction from a coordinating institution to a command structure that is full of entity power and responsibility, similar to disaster management function of Japan's Cabinet Office. Because of the expansion of function, China National Committee for Disaster Reduction should be renamed accordingly, and it can be renamed as "National Emergency Responding Committee" based on *Law of the People's Republic of China on Emergency Response*. Emergencies under the Conceptual framework of large-scale disasters are incorporated into the managed object that should strengthen business integration of relevant disaster prevention and mitigation departments through innovative mechanism, and integrate various types of disaster relief resources. This will not only help to improve implementation capacity of government in handling emergencies, but also help to promote the allocation efficiency of various types of disaster relief resources.

4.3 Strengthen disaster management and implication role of local government

In general, disasters always happen in certain areas, so disaster emergency management should involve governments at all levels.

Especially for local governments, they must figure out their specific tasks in disaster prevention and mitigation, and improve its efficiency and effectiveness.

Firstly, strengthen the city district and rural township government as the first line of function implementation in disaster prevention and reduction. Take into account the regional characteristics of disaster prevention and mitigation work. Only when they are effectively combined with local disaster prevention and reduction resources, and good daily preparation is made for all disaster prevention and mitigation, can we mobilize quickly in the event of a disaster rescue. Especially for multi-spot, large-scale disaster, it is more important to use local disaster relief resources, in which local government plays a significant role. The author proposes integrating relevant departments, building emergency management department in local government, inviting department heads into the Party Standing Committee. Its management scope includes responding not only to natural disasters, but also to incidents of social groups, as long as the events related to public security. Secondly, administrative regions are almost starting point of implementation of a variety of disaster prevention and mitigation planning because of the segmentation of administrative regions. However, disasters are often beyond administrative regions. Controlling and allocating resources is not easy when the disaster information is difficult to clearly transmit, so it is important to set up cross-regional joint mechanism to coordinate the disaster prevention and mitigation act of different local governments, thus to improve the efficiency of disaster relief. Finally, the central government should establish supervision and evaluation system for local governments' disaster prevention, review the disaster prevention and mitigation planning and operability of response measures in every region, and also urge local governments to implement disaster mitigation and preparedness. When a disaster occurs, it provides support for disaster evaluation and emergency rescue mechanism, sets clear disaster support level, and avoids ineffective scheduling of disaster relief resources.

4.4 Establish a disaster prevention and mitigation network with smooth flow of information

China should establish unified model of structure for command and execution as well as standardized management principles, which can contribute to the integration, coordination, command, deployment and scheduling for various types of disaster relief resources in disaster regions to ensure that on-site disaster command be clear and smooth, and to increase the implementation effect of disaster prevention and mitigation. With the benefits from the developed information industry, e-business data and network level of relevant Chinese governments at all levels are relatively high. The key problem now is that governments at all levels are confined to their own concept of departments, and have not established smooth communication channels among different departments.

To begin with, we should establish a unified disaster management system on the basis of our country's actual situation, regard it as a standardized platform for disaster response, and strengthen united command in disaster prevention and mitigation. Moreover, we should develop information platform in multi-level modular training, reproduce the simulation of disaster scenarios, take joint training depending on the disaster scale, scenario simulation conditions, evolution of the disaster and identity of participants, by which we can strengthen commanders' ability to judge the situation, and accumulate experience in disaster response by systematic work and experiential study. Last but not the least, government departments should set up their own real-time horizontal and vertical linkages based on information network. At the same time, they should get access to the Internet so that ordinary people can receive real information on disaster prevention and reduction timely, thus building a disaster prevention and mitigation network with smooth flow of information that ordinary people can share with governments at all levels.

4.5 Strengthen social groups' role in disaster prevention and mitigation

When a large-scale disaster happens, it is often difficult to do everything with the government's disaster relief efforts alone, so the integration of civic and charitable organizations involving in disaster relief work is necessary, which can help to promote public self-defense capability, and to make full use of civil disaster relief capacity. China's disaster relief status can be summarized as "emphasizing on large-scale disasters while pay little attention to small ones, emphasizing on government's disaster reduction work while pay little attention to the society's, emphasizing on post-disaster rescue while pay little attention to pre-disaster prevention, emphasizing on direct loss of life and property while pay little attention to manufacturing and social loss" [1]. However, disaster relief activities that government ignore or are not taken into account should be made up for through social organizations' participation. It is the scope for NGOs. But it is not clear to government that how to promote the breadth and depth of civil participation. Meanwhile, many NGOs still do not understand what specific resources government possesses in disaster relief activities. So we should set up communication platform to strengthen the contact between government and social groups.

Firstly, we should establish a special department to communicate with social organizations in implementing agencies of China National Committee for Disaster Reduction, and its duty is not only to promote communication between government departments and social organizations, but more importantly is to coordinate and manage social organizations to participate in disaster prevention and relief for efficient resources allocation. On the one hand, it can be conducive to exchanging information timely between social organizations and the government to avoid blind using of relief resources, duplication or waste; on the other hand, it can also inform social organizations of disaster relief information and allocation of official resources to improve

---

[1] Shen Jianli. *The Ministry of Civil Affairs Pushes Cooperation Between Officials and Citizens, NGOs Intend to Build a Shared Network*. 21st Century Economic Report, April 6, 2011.

disaster relief efficiency. Secondly, NGOs' network platform in disaster prevention and relief should be established among social organizations. As a mode of hub mechanism and communication of information resources we should establish a more prestigious local NGO in all provinces, and be responsible for disaster prevention training work. Then, we can choose several executive agencies to take specific project implementation by biding.

4.6 Strengthen public participation in disaster prevention and mitigation

Firstly, we should develop disaster prevention awareness of the whole nation and pay attention to personal implementation experience. Kobe earthquake in Japan in 1995 resulted in serious destruction, after which Japanese government designated September 1st in each year as "Disaster Prevention Day". In addition to continuous expansion of the comprehensive training, Japanese government focuses on public participation in disaster drills, which are exercised in real buildings so that residents can have practical field experience. It also establishes the real sectors, and provides facilities for people to practice. At present, China still spread prevention methods via text, media and so on. However, ordinary people rarely have opportunities to practice. Therefore, China should learn from Japan to improve people's knowledge of disaster prevention and relief by encouraging them to exercise practical implementation. Secondly, we should strengthen civilian disaster prevention organizations. The campaign of "Individual Self-aid Capacity in Disaster" should be promoted in each family, workplace, school and community to improve the responding capacity of the public.

References:

[1] Tong Xing, Zhang Haibo. Disaster Management Framework Based on Analysis of China's Problems[J]. Social Sciences in China, 2010, No. 1.

[2] Wang Dexun. The Implications of Japan's Earthquake Disaster Mitigation [J]. Journal of Academy of Social Sciences Institute, 2008, No. 42.

[3] Yuan Yi. Japan's Legal System of Disaster Management[J]. Disaster Reduction in China, 2004, No. 12.

[4] Yuan Yi. Japan's Administrative System and Reduction Plans for Disaster Management[J]. Disaster Reduction in China, 2004, No. 12.

[5] Li Jun, Nie Yingde. Japan's Disaster Information System and Its Operation: Experience and Implications[J]. Around Southeast Asia, 2009, No. 2.

[6] Xiong Guanghua, Wu Xiuguang, Ye Junxing et al. The Evolution of Taiwan's Disaster Prevention and Relief System: The 2010 Cross-straits Public Governance Forum —Public Administration, Disaster Response and Crisis Management. Social Science Literature Publishing House, 2011.

[7] Shen Jianli. The Ministry of Civil Affairs Pushes Cooperation Between Officials and Citizens, NGOs Intend to Build a Shared Network[J]. 21st Century Economic Report, April 6, 2011.

[8] Lu Cong, Wei Yiming, Fan Ying, Xu Weixuan. Economic Impact of Disasters on the Quantitative Analysis Model and Its Application[J]. Journal of Natural Disasters, 2002, No. 4.

[9] Zhao Yandong. Social Capital and Post-disaster Recovery[J]. Sociological Studies, 2007, No. 5.

[10] Cabinet Office. 災害プレースタッフ研修契約情報, Heisei, 2009(21).

(**Translated by Cai Zhihai**)

A Study on the Influence of Natural Disaster on Clustering Contiguous Special Poverty-stricken Communities[*]

—Taking the Wuling Mountainous Area as an Example

Zhang Dawei

Abstract: The natural disaster can have an impact on poverty. Disaster-risk, vulnerability, feasible ability and poverty have the intense mutual construction in clustering contiguous special poverty-stricken communities. The analytic frame which generated from the four core essential factors can well explain the influence of natural disaster on poverty. From the perspective of vulnerability and history, feasible ability and reality, poverty and future, we can see that compared with

---

[*] One of the phased achievements of Chinese Post-Doctorate Scientific Fund Project "the contiguous development research on the special-type impoverished area in Wuling mountainous area"(201000480919), the National Social Science Fund Project "the research on the integration construction of community public service system in the process of urban and rural overall planning advancement"(09CZZ025), "The poverty alleviation and development research in concentrated and contiguous special-type difficult area (WuLing mountainous area)"(201000402)of Huazhong Normal University. The data in this paper came from "the poverty alleviation strategic research in concentrated and contiguous special difficult area (WuLing mountainous area)"of baseline survey group in which the author participated. This group is composed by the International Poverty Relief Center of China (IPRCC), Germany International Cooperation Agency(GIZ), the Sociological Institute of Central China Normal University. My hearty thanks goes to all the group members.

Zhang Dawei, a sociology postdoctoral and researcher of Social Development and Policy Research Center in Huazhong Normal University.

the common communities, the risk caused by natural disaster has a bigger influence on difficult communities. Through the analysis of the Wuling mountainous area, we can discover that the natural disaster happens in high frequency and large numbers in clustering contiguous special poverty-stricken communities; the damage of natural disaster is wide and deep; natural disaster and poverty have coincident property and consistency; there are relative alternating property and cyclicity between natural disasters, vulnerability, feasible ability, poverty and so on. Strengthening the management of disaster risk in concentrated and contiguous special poverty-stricken communities can be significant and imperative.

Key words: Natural Disaster; Poverty; Poverty-stricken Community; Wuling Mountainous Area; Clustering Contiguous

The research on the impact of natural disaster on poverty is a new project. During the Eleventh Five-year Plan period, earthquake, flood, drought and debris flow and the like occurred frequently, rural productivity and lives faced serious challenges in our country. Natural disasters are often accompanied with poverty, especially in clustering contiguous special poverty-stricken communities, and the intervention of disaster-risk made the poverty alleviation and development more complex in these communities. The national poverty alleviation conference held at the end of 2010 definitely pointed out that our country would regard the clustering contiguous special poverty-stricken areas as the main battlefield, and would pay more attention to solve the poverty problem in these areas in the next 10 years. On March 16, 2011, the Xinhua News Agency published the Twelfth Five-Year Plan (2011—2015) which not only pointed out that "to strengthen the monitoring, early warning and prevention of extreme weather and climate events, and to improve the ability of defending and reducing natural disasters", but also emphasized "to implement the crucial poverty alleviation and development projects in the southern boundary areas, the eastern edge of the Qinghai-Tibet Plateau, Wuling mountainous areas and other clustering contiguous special poverty-stricken areas". In view of this, introducing natural disasters into this special field and inspecting its influences on poverty, researching the general rules of how the natural disasters and poverty occurr, and discussing the ways and means of the disaster-risk management and poverty alleviation are of great value.

1. The Analytic Frame of Disaster Risk and Community Poverty

This article will establish an analytic frame of how the disaster risk affects the poverty-stricken communities, namely the analytic frame of disaster risk and community poverty. It is based on the theory of the essential factors' mutual construction 'At first' it will look for the accurate core factors or the essential concepts in the interaction between disaster and poverty, then it tries to establish the related model or analytic frame on the basis of defining the connotation of various essential factors.

1.1 Concept of the core elements

Disaster-risk, poverty, vulnerability and feasible ability are the core elements or the essential concepts in the process of natural disasters affecting poverty. Firstly, the disaster-risk. Generally speaking, risk refers to a combination of the possibility and consequence resulted from some specific dangerous situations. The narrow definition of risk emphasizes the uncertainty (the result must be bad). The generalized definition emphasizes the uncertainty of loss (the result may be disadvantageous or possibly advantageous or compromise). Disaster-risk is more likely to bring the negative consequences, "They pose the inevitable threat to plants, animals and the human lives"[1]. Secondly, poverty. The concept of poverty is very complex. The United Nations Development Program defined poverty as "it is not just the low-

---

[1] Ullrich Beck. *Risk Society*. Translated by by He Bowen. Nanjing: Yilin Press, 2004, P. 7.

incomers, but also refers to those lack of medical treatment and education, deprivation of the right of being educated and communication, the deprivation of human rights and the political rights, lack of dignity, confidence and self-esteem" in The Human Development Report in 1996. The Nobel Prize winner Amartya Sen believed that "poverty is not merely the problem of the poor with a low income, but also the problem of poor people's lack of the ability of obtaining and enjoying their normal life, or the true meaning of poverty is that impoverished population has poor ability or lacks opportunity to earn their income". Taking all these viewpoints into consideration, this article defined poverty as the following: People cannot get enough income to maintain basic daily expense, or obtain the basic survival resources; the human resources and the social participation resources are less than the lowest living standard identified by the society. It generally includes the insufficiency of material, economy, ability, right and so on. Thirdly, vulnerability. The concept of vulnerability was the evolution of the ecological concept, and it can refer to the ecological vulnerability and the impoverished people's vulnerability when this concept is integrated into the reasearch of poverty. The vulnerability of the impoverished subject mainly refers to each kind of possibilities which cause impoverished risk the community, family and individual will face. The higher the vulnerability is, the more possible the subject becomes impoverished. Generally speaking, the feasible ability is weakened when the vulnerability strengthens. Fourthly, feasible ability. Amartya Sen thought that, "one person's feasible ability means the combination of all kinds of possible functional activities and the realization of these activities. Therefore, the feasible ability is a kind of freedom, a substantive freedom which can realize all kinds of possible combination of functional activities"[1]. This paper believes that, this kind of freedom, including the economic capacity, political ability, social ability, cultural ability

---

[1] Amartya K. Sen. Development as Freedom. Translated by Ren Ze, Yu Zhen (ed.). Beijing: Renmin University of China Press, 2002, P. 62.

and so on, which can make people live above the lowest survival conditions and can change people's status in reality. Moreover, this paper also regards community and family as the main body of feasible ability.

1.2 The analysis framework of the mutual construction of essential factors

Disaster-risk, vulnerability, feasible ability and poverty have the intrinsic mutual construction of essential factors. ① This kind of mutual construction is revealed more sufficiently in clustering contiguous special poverty-stricken communities, The natural disaster is a kind of risk, and the risk can strengthen the vulnerability of the community, family and individual, weaken their feasible ability, thus causing poverty or returning to poverty. If we regard the poverty-stricken communities as the special field to study the influence of natural disaster risks on impoverished communities, we can discover that the risks caused by natural disasters have greater influence on poverty-stricken communities and people than common communities: First, from the perspective of vulnerability and history, the poverty-stricken communities are more vulnerable and natural disasters are easier to occur in these places, which increase the possibility of deepening poverty. As some scholars said, in history, the poorest people generally live in the most dangerous places, and the poverty-stricken villages are also located in the remote areas far away from the major roads, and the ecological geography and the natural environment are bad there. These places have more vulnerability, and the frequency of natural disasters is relatively higher. Once the natural disasters happen, the poverty degree will deepen inevitably. Secondly, from the perspective of feasible ability and reality, in the three stages namely, before natural disasters happen,

---

① The mutual construction of essential factors in this article was inspired by social mutual-construction put forward by Zheng Hangsheng. Reference from Zheng Hangsheng, Yang Min: Social Mutual-Construction: the New Exploration of Sociology Theory with Chinese Characteristics under Worldly Wisdom: "The research of Relation between Individual and Society" in Contemporary China. Renmin University of China Press, 2010.

during the natural disasters and after the disasters, the poverty-stricken communities and the people showed an insufficiency of feasible ability in disaster prevention, disaster reduction and reconstruction, which also intensified the poverty degree. On the one hand, before natural disasters happen, the disaster prevention ability of the poverty-stricken communities and the people are relatively poor, which is manifested in the single economic structure, low cultural quality, backwardness of mind, weak material foundation, lack of self-protection skills, and insufficient feasible ability of preventing disaster. On the other hand, when a natural disaster occurs, the disaster prevention ability of the poverty-stricken communities and the people are relatively poor, which is manifested in the bad quality of housing, the weak resistance ability, the insufficiency of knowledge of avoiding disaster and so on. On the third hand, after natural disasters, the reconstruction ability of the poverty-stricken communities and people is relatively insufficient, which is manifested in the great damage on the original foundation, the insufficiency of capital and material. The inconvenient access to outside help also adds to the long and difficult process of recovery and reconstruction[1]. Thirdly, from the perspective of poverty and future, these poverty-stricken communities becomes more impoverished after natural disasters, and their vulnerability become greater and their feasible ability becomes lower, which easily incurs the coming-again of natural disasters. The poverty-stricken communities become poorer because of this vicious circle.

2. The Interpretation of Wuling Mountainous Area—Poverty-stricken Communities

The following section will carry on explanation from the research object and the research method. The realistic condition of natural

---

[1] Huang Chengwei & Lu Hanwen. *The Reconstruction Process and Challenges of Impoverished Village after the Wenchuan Earthquake*. Beijing: Social Sciences Academic Press, 2011, P. 5.

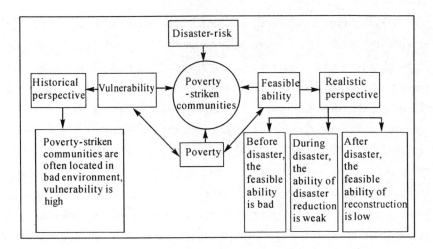

Figure 1 The analysis frame of mutual construction of disaster-risk, vulnerability, feasible ability and poverty

disasters in the Wuling mountainous area—poverty-stricken communities, the influences of natural disaster on Wuling mountainous area's clustering contiguous special poverty-stricken communities and so on.

2.1 The survey and research method of Wuling mountainous area-poverty-stricken communities and research methods

The Wuling Mountain is located in the conjunction of Chongqing, Hubei, Hunan and Guizhou. The Wuling mountainous area is also called the Wuling contiguous mountainous area. Surrounded by the Wuling mountain, it is a strategic connective region of the development of the western region and the rise of the central China, and it is also one of the 18 contiguously impoverished areas where our country has given the key support. The 18 contiguously impoverished areas are the collection of former revolutionary areas, minority regions, boundary areas, impoverished areas and mountainous areas. There are 56 counties in Wuling mountainous areas, among which 30 counties are state-level poverty-stricken counties. It is one of the most accumulated poverty-stricken areas. Wuling mountainous area is primarily of Karst landscape with an total area of 100,000 square kilometers, and the total population is more than 20 million, and it has one of the largest minority assembling communities over the province border, including more than 30 nationalities such as Tujia, Miao, Dong, Bai, Hui, Yao and so on, more than 12 million people. ① Wuling mountainous area is truly a clustering contiguous special poverty-stricken community with the features of the adjacent mountainous poverty, the assembling national minorities and the fragile natural ecological environment and so on. At the end of 2010, the strategic research group(in which the author was included) of poverty reduction in the clustering contiguous special poverty-stricken areas(Wuling mountainous area) was divided into four study groups, and each group went separately to Chongqing City, Hubei Province, Hunan Province and Guizhou Province to carry out the study of poverty alleviation and development concerning Wuling mountainous area. We provided and collected acquisition tables from 149 poverty-stricken communities foundation data and 698 effective villager questionnaires by using date collection, case interview, group discussion, field observation and other methods. The following paragraphs will take the data of this investigation as a foundation to carry on the case analysis of mutual construction, regularity of disaster-risk, vulnerability, feasible ability and poverty, and explain the rules by which the natural disasters affect poverty-stricken areas.

2.2 The natural disaster situation of Wuling mountainous area—poverty-stricken communities

The natural disasters that happened in Wuling mountainous area poverty-stricken communities has the features of multiplicity and high frequency. On the one hand, the natural disasters in this area have multiformity. The common natural disasters in this area are drought, frost, wind, mountain torrent, mud-rock flow, landslide, subsidence

---

① Feng Quanguang. *Mountainous region economy synergistic development among the edges of administrative areas under the geo-economic regional perspective : Taking the Wuling Mountainous Regions in combination of Chongqing, Guizhou, Hubei and Hunan as examples.* Journal of Mountain Science, 2009, Vol. 2, P. 170-171.

and so on. The investigation demonstrated that, in recent 5 years, various kinds of disasters occurred in the poverty-stricken communities, in the 124 effective sample communities. There has been 366 floods, 290 insect-caused disasters, 281 wind-caused disasters, 169 landslides triggered by floods, 121 ice hail disasters, 117 frost colds, 66 brushfires and moreover 28 other type disasters. On the other hand, the Wuling mountainous area poverty-stricken communities' natural disasters have a high frequency. The flood disaster, insect-caused disaster, wind-caused disaster, drought happened frequently in this region. The investigation demonstrated that, in recent 5 years, each community had suffered 3 floods, 2.3 insect-caused disasters, 2.3 wind-caused disasters, 2.3 droughts, 1.4 mountain torrents, 1 ice hail disaster, 0.9 frost cold and 0.5 fore stfires. The natural disasters happen frequently in various areas. For example, according to the statistics, during the previous 95 years in 20th century, flood happened in 74 years in poverty-stricken communities in Xiangxi Prefecture of Hunan Province. Flood covered all the Xiangxi Prefecture in 35 years. Extremely big flood happened during 10 years; flood happened in 33 years from 1958 to 1995. That is to say, flood happened for 9 years in each decade. The frequency of drought is 73%-92%, the severe drought happens every 2-3 years, and the frequency accelerates year by year. ①

The influence of natural disasters on Wuling mountainous area is deep and wide. On the one hand, serious natural disaster affected broad areas. The investigation demonstrated that, in the recent 5 years, different types of natural disasters happened in poverty-stricken communities, including drought, flood, ice hail disaster, insect-caused disaster, wind-caused disaster, frost cold, mud-flow disaster, snow disaster, electricity disaster, fire and so on. Taking a look at the affected villages, we could see that these disasters affected respectively 79, 78, 33, 29, 28, 17, 5, 2, 1 and 1 village of the 124 villages, in

---

① Bai Jinxiang. *Study on Economic Development of Mountain Fastness*. The Ethnic Publishing House, 2006, P. 47.

which a lot of villages suffered serious disasters. Moreover, in the recent 5 years, 69 villages have experienced severe disasters three times, accounting for 42% of the total number of villages. 90 villages experienced severe disasters two times, making up 54.9% of the total number of villages, and 102 villages experienced severe disaster one time, taking up 61.8% of the total number of villages. On the other hand, the loss caused by serious natural disaster is heavy. The investigation demonstrated that, the average damaging area of each seriously affected village was about 500 mu, and the direct economic loss has surpassed ¥ 130,000. Moreover, the rate of crop loss caused by diseases and insects in Wuling mountainous areas poverty-stricken communities was as follows: The corn, rape and tobacco were between 5%-12%, rice was between 9.48%-12.39%, the citrus fruits reached as high as 30%, and the harvest suffered seriously.

2.3 The influence of natural disaster on the Wuling mountainous area poverty-stricken communities

(1) From the perspective of vulnerability and history

From the perspective of vulnerability, there are close inner links between ecological fragility and disaster, between disaster and poverty. Ecological fragility easily leads to natural disasters, thus it is easy to cause poverty. Looking back at the development of social history, the poor often live in the place which the ecology and the environment are fragile, and the poverty-stricken community often has high vulnerability. The combination of these two factors leads to the following results, the poverty-stricken community has high vulnerability, and the higher vulnerability is easier to cause natural disasters; the impoverished community, is easier to have natural disasters, and natural disasters increase the poverty-stricken communities' poverty degree. This is fully reflected in clustering contiguous special poverty-stricken communities in Wuling mountainous area.

The poverty of the poverty-stricken communities in Wuling mountainous area can not be separated from its ecological vulnerability.

A Study on the Influence of Natural Disaster on Clustering Contiguous Special Poverty-stricken Communities

The living environment is bad, and there are many remote mountains and alpine zones. The Wuling mountain is a fold mountain, which has a series of folds and fractures, and many of them are of Karst landscape, presenting multistage altiplanation surface with the height ranging from 350 meters to 1200 meters. Case study demonstrated that, poverty-stricken communities are mainly located in the remote mountains or half way up the mountains, this kind of communities occupied 89.3%. The poverty-stricken communities of the Wuling are mainly located in the remote places far away from the major artery roads, and they are lack of the sufficient means of transportation. The statistical analysis of the sample peasant households demonstrated that, the average distance from the peasants' homes to the town is about 9.6 kilometers, 5% villagers are more than 25 kilometers away from the town, and there are also the mountain roads. The one-way time of getting to the town is generally 100 minutes, and the longest time is 600 minutes. The natural endowment of the poverty-stricken community in Wuling mountainous area is bad. In this area, mountainous area and hill land account for above 95%. There is little contiguous cultivatable land, however the dispersive slope farmland and terraces over 15° are numerous, and the soil layer is also thin with low output, which altogether cause the low productivity of cultivatable land and the weak land supporting capacity[1]. Once the natural disasters occur, villagers will be threatened with food shortage. In the investigation of these four provinces (cities), some villagers said frankly, "we are inevitably suffering from natural disasters in such environment, and we cannot but be impoverished". Each area reflected that ecological fragility increases the possibility of natural disasters, thus it restricted the progress of poverty alleviation and development.

Judging from this, the poverty-stricken communities of the Wuling mountainous area are mainly located in the disaster-prone area which has

---

[1] Quan Fang. *The thought of solving the poverty of multinational area in Wuling mountain.* Journal of Guizhou Provincial Committee Party's School of CPC, 2008, Vol 6, P. 95.

strong ecological and environmental vulnerability. Therefore, the frequency of natural disaster is high in poverty-stricken communities. Once a natural disaster occurs, it will generally cause the failure of crops, or no crop, the possibility of deepening poverty gets enhances.

(2) From the perspective of feasible ability and reality

The poverty-stricken communities' vulnerability is also shown by its insufficient feasible ability of changing present situation. From the perspective of reality, natural disaster has three stages, which are before the occurrence, during the occurrence and after the occurrence, yet the poverty-stricken community's feasible ability during all of the stages is insufficient: before the natural disaster, the capacity of disaster prevention is bad; during the natural disaster, the ability of disaster relief is weak; and after the disaster, the feasible ability of reconstruction is weak. It is also easier to make the poverty-stricken communities poorer and the poverty degree deepens.

Before the natural disaster, the poverty-stricken community's weak inner feasible ability of preventing disaster, which are mainly manifested by shortage of financial capital, insufficiency of material capital, deficient human capital, weak social capital, shortage of disaster-avoiding knowledge, poor disaster information, weak village-level organization and so on. The following paragraph will carry on the analysis from the shortage of financial capital, material capital, human capital and weak social capital, etc.

First, the shortage of financial capital and the insufficient material capital. The agricultural structure of the poverty-stricken communities of Wuling mountainous area is unbalanced. The crop here is monotonous and has low output and profit. The investigation demonstrated that, in 149 sampling villages, rice and corn are the main plants in spring. Among 115 villages, there are 68,586 mu of paddy rice, accounting for 40.4% of the total crop planting area, and 82,073 mu of corn, accounting for 48.4% of the total crop planting area. The two crops' planting areas take up about 89.2% of the total area. According to panel discussion and interview, although there are many plants, only about 2 kinds of plants can bring income, which are mainly the traditional grain crops with low output and profit. The cash crops are few. In addition, in 149 sampling villages, nearly half of villages there can only cultivate one kind of crop. Among the 82 villages which are able to cultivate two crops, the main crops are wheat, potato, sweet potato in autumn. The investigation demonstrates that, these crops often depend on weather; their output is low, most of which are for self-use with few commodity tradings. In addition to the single variety of food crops, the industrial crops are also in monotonous variety and of small planting area. Therefore, the poverty-stricken communities of Wuling mountainous area have monotonous economic structure and are easier to suffer from disasters; and the insufficient goods and materials cause the weak feasible ability of preventing disaster. Second, the deficient human capital and the weak social capital. On the one hand, most villagers' educational level is low. The survey shows that 92.2% interviewees only receive secondary education or just elementary education. Among them, there are only 7 people who receive junior college education and the above, only accounting for 1% of the total number. 47 people receive high school education and the above, accounting for 6.8% of the total number, 197 people receive secondary education, accounting for 28.6% of the total number. 328 people receive the elementary education or below, accounting for 47.5% of the total number. In addition, 118 people never receive education, accounting for 17.2% of the total number. The investigation indicates that, insufficient education also restricts the ability of accepting technical knowledge and causes the shortage of ability to avoid disasters. At the same time, villagers' mind is backward as a whole, and they have no innovative and enterprising spirit, being content with the status quo. The backwardness of thought will affect the feasible ability. On the other hand, the social network is relatively narrow. When asked about who you can seek for help when you meet difficulties, the answers respectively are: relative occupies 69%, neighbor occupies 40.5%, family occupies 33.8%, friend occupies 33.7%, the government occupies 25.2%. The role of the government is extremely limited in villagers' daily life, and the peasants are more likely to seek for help from the lower levels and horizontal human relationships, lack of communication with the higher level relationship leads to deficiency in the vertical linkage of social capital. This shows the low educational level of the villagers living in the poverty-stricken communities in Wuling mountainous area, the deficient human capital; frail social capital and the weak feasible ability of preventing disasters.

When natural disaster occurs, the inner feasible ability of reducing disaster of the poverty-stricken communities is weak, which are manifested by the poor quality of building, vulnerable economic basis, low emergency capability, weak organizing ability and so on. The following section will mainly analyze from the bad quality of building and the weak economic basis, etc.

Firstly, as the most important property, the antique houses with simple structure and poor ability of preventing disasters will suffer from huge damage once the earthquake, wind disaster and snow disaster occur. In light of the housing structure, the houses which are regarded as the status symbol or as the most important material of living foundation are comparatively backward: For one thing, most houses were built decades ago. They are generally old and traditional. The houses are constructed mainly around 1980s, 50% houses are constructed before 1982, and most houses were built several decades ago or more. The oldest house has more than 200 years of history. For the other thing, the structure is mostly made of wood and soil. The statistics demonstrate that 68.9% of the interviewed farmers' house structure are built by wood and soil, the proportion of the houses made of concrete and brick is less than 30% and the houses even shake when the crowd are passing. Therefore, the house quality of the poverty-stricken communities in Wuling mountainous area is bad, and the feasible ability of evading disaster is weak. Secondly, because of the single economic structure the source of income is unstable, and the capability of dealing with emergency is frail. The statistical analysis discovers that, the difference between the annual average income and the annual disbursement is less than ￥ 300 in the family with no family member woring in other places. That is to say, the annual income and the annual disbursement are basically balanced, the surplus is below ￥ 300. Therefore, the capability of dealing with emergency of the farmer's family with no one working outside is very weak. In addition, generally speaking, the economic condition of the families with members working outside is better than that of the farmer's family with no family member working outside. According to the interview, because of their low education and bad health, most of the farmers working outside are engaged in drudge work which have huge risk and unstability, and they are often unemployed. So even some members in the family work outside, its ability of dealing with various emergencies such as natural disasters is also frail. Thus it can be seen that, the economic foundation of the poverty-stricken communities in Wuling mountainous area is weak, and the feasible ability of reducing disasters is weak too.

After the natural disaster, the inner reconstruction feasible ability is low, which are mainly manifested by insufficient fund and commodities, difficulty in entering of aid from outside, the great damage to original foundation, difficulty in reconstruction, long time of recovery and so on. The following paragraphs will analyze from the insufficient fund and commodity and the difficulty in entering of aid from outside and so on.

First, the economic basis is weak, and the fund and commodities are insufficient and unavailable after the disaster. On the one hand, the per capita income is low in poverty-stricken communities. According to the survey, in the 8 sample counties, the average capita income of farmers of Xiushan county in chongqing city is only ￥ 3,447, which is the highest income in 2009, the average per capita income of farmers of Yinjiang county in Guizhou province is only ￥ 2,610 which is the lowest income, the income of Xuan'en county of Hubei province is ￥ 2,804, the income of Xianfeng county of Hubei province is ￥ 2,806, the income of Si'nan county of Guizhou province is ￥ 2,839, the income of Luxi county of Hunan province is ￥ 2,839, the income of Fenghuang county of Hunan province is ￥ 3,145. The per capita income of these counties are between the highest income and the lowest income, and all of them are lower than the national average level of ￥ 5,153. On the other hand, the farmer's annual expense is large and the surplus is little. Although the income of the farmer is not high, the disbursement is actually high. The average surplus of each peasant household was ￥ 2,770.7 in 2010, which was hard for the family to support 5 people. Moreover, farmers often borrow money. According to the investigation, more than half of the families have loan, and the highest loan amounts to ￥ 130,000. Therefore, due to the weak economic basis of the poverty-stricken communities in Wuling mountainous area, once a disaster occurs, the available fund and commodities are not enough. The reconstruction feasible ability is insufficient too, thus reconstruction is in difficulty. Secondly, because of the weak infrastructure, it's difficult for outside aid to enter these places after the disasters. Firstly, the roads are inconvenient. The roads in the poverty-stricken communities of Wuling mountainous area have no access to the every group even most farmer families. The investigation demonstrated that, 2 villager groups in each village have no access to road, and unexpectedly 18 villager groups have no access to road in a village; on average each village has 78 farmer households have no access to the road. There is even a village with 3,616 households which have no access to road. Obviously, the transportation condition is so bad in this area. Secondly, the low constructive level of electricity, communication and other service facilities. In terms of electricity, the investigation demonstrates that, although the coverage ratio of electrical network in sample village is close to 100%, according to on-site interview, due to the electricity is often affected by natural disasters or human factors, the using of electricity is not guaranteed in impoverished villages, and the price is one time higher than that of the ordinary electrovalence. In terms of telephone use, generally speaking, each village has 50 telephones, accounting for 12.3% of the total families; there are 200 mobile phones which accounts for 12.5% of the total villagers; the rate of telephone use is low, the signal is bad and often interrupted. Thus it can be seen that, the infrastructure of poverty-stricken communities in Wuling mountainous area is bad. Once the natural disaster occurs, these communities will suffer a lot, and it is hard to contact with the outside world. At the same time the external aid is also difficult to enter the area, and the feasible ability of reconstruction in the poverty-stricken communities is insufficient and reconstruction is in great difficulty.

(3) The perspective of poverty and future

The natural ecological vulnerability is high in Wuling mountainous area where the poverty-stricken communities are located, and it increases the occurrence of natural disasters. Once the natural disasters happen in the poverty-stricken communities of Wuling mountainous area, it would further enlarge the vulnerability and reduce the feasible ability of impoverished villages. If people don't carry out the prompt measures of recovery and reconstruction in the poverty-stricken communities which suffer from disasters, or adopt the effective disaster risk management, the poverty degree will deepen in these communities, and the more vulnerability and lower feasible ability will lead to another disaster. In such a vicious circle, the poverty-stricken communities will become even more impoverished. In this investigation, we discover that, disasters not only deepen the poverty of the poverty-stricken communities, but also make the farmers who have gotten rid of poverty after the last disaster return to poverty again. In the interview, we find that the disaster is the production of "the small regional" factors, which can make the poverty-stricken communities that suffer from the disasters return to poverty again easily. Previous experience indicates that, the farmers who have gradually escaped from poverty after the previous disaster will return to poverty once again due to the lack of the effective aid of disaster risk management and self-service. Therefore, from the perspective of future, the damaged poverty-stricken communities which are lack of the disaster risk management are easier to incur disasters, and easier to aggravate poverty. The following examples will verify this judgment well.

Although the difficult communities in Wuling mountainous area often suffer from natural disasters, some villagers still recover gradually and tend to get rid of poverty between the two serious natural disasters. Unfortunately, they are likely to encounter the attacks of natural disasters again because of insufficient effective disaster risk management aid, self-help mechanism and measures, which sometimes make these villagers too difficult to live on, and the phenomenon of returning to poverty is prominent. For example, from 2010 to 2011, in Youyang county of Chongqing city, there were 129,873 people who needed help to survive the winter and spring because of disasters, accounting for 18.7% of the total population of agriculture. Among them, there were 107,632 people in need of grain ration, 12,680 people in need of water, 3,465 people in need of clothing, 4,682 people in need of heat, 1,414 people in need of medical rescue. Thus it can be seen that the disaster was so serious that the villagers who are in face of the problem of grain ration occupied 82.9% of total affected population. What is unexpected is that, half of the affected villagers are the people who gradually get rid of poverty after the last natural disaster. After the survey of 9 disaster-hit households which face this kind situation, we discovered that, there are 8 households whose grain crops production has decreased, the worst underproduction is 33% and the least underproduction is 8%, only one household's grain production maintains the same level with the last year's, which had already reduced massively last year because of the disaster. In these 9 families, because of the influences of disasters and the insufficiency of grain ration, the family with the most grain ration can only maintain 220 days, and the family with the least grain ration can only survive 120 days. The phenomenon of returning to poverty is serious(see Table 1).

From the above analysis, we can see that it is necessary to strengthen the disaster risk management and pay attention to the poverty-stricken communities' future development after natural disasters. Because there is a logic: If the poverty-stricken communities'

A Study on the Influence of Natural Disaster on Clustering Contiguous Special Poverty-stricken Communities

Table 1 The situation of the villagers who return to poverty because of the disaster after having escaped from poverty

Organization's name	House-holder's name	Family population	The harvest in 2009(catty)			The harvest in 2010(catty)			Year-on-year Underproduction(%)	Ration maintaining days
			Rice	Corn	Sweet potato	Rice	Corn	Sweet potato		
Pingdiba village of Heishui town	Tan xx	4	600	500	0	500	400	0	8	220
Pingdiba village of Heishui town	Guo xx	5	700	500	0	700	500	0	0	190
Tiancang village of Maoba town	Li xx	4	400	600	700	400	400	700	11	120
Xishahe village of Maoba town	Zeng xx	6	800	800	1200	700	700	1100	10	150
Xishahe village of Maoba town	Yang xx	4	500	500	1000	500	500	900	5	120
Huijia village of Dingshi town	Li xx	3	500	700	300	400	500	200	27	220
Jinshan village of Dingshi town	Ran xx	2	0	600	0	0	400	0	33	210
Sanxikou village of Dingshi town	Ran xx	4	600	600	500	400	500	400	24	210

disaster risk management is deficient, the temporary superficial restoration is accomplished, higher vulnerability and lower feasible ability would be generated, and these communities may suffer again. Because of this vicious circle, the poverty-stricken communities will become more impoverished.

3. The Basic Judgment of Theory Presupposition-Experience Study

The poverty problem would become more complex after natural disasters happen in poverty-stricken communities. If we regard the poverty-stricken community as the special field of researching the influence of disaster risk on community poverty, we can see that, from the perspective of the vulnerability and history, the inherent vulnerability of poverty-stricken communities makes it easier to cause disaster; from the perspective of feasible ability and reality, the tendency of insufficiency of the feasible ability during the three stages of disasters is more likely to deepen poverty; from the perspective of poverty and future, the poverty-stricken communities that suffer from disasters will become more fragile due to the insufficiency of disaster risk management, and it will become more impoverished finally. In summary, compared with the common communities, the influences of risk which caused by disasters on the poverty-stricken communities and the impoverished crowd are greater. Through the research on the influence of disaster in clustering contiguous special poverty-stricken communities in Wuling mountainous area on the poverty issue, the following basic judgments can be attained:

Firstly, disaster-risk, vulnerability, feasible ability and poverty have an intrinsic mutual construction of essential factors. This mutual construction is reflected more fully in clustering contiguous special poverty-stricken communities. Secondly, the influence of natural disaster on clustering contiguous special poverty-stricken communities is deep and wide. The clustering contiguous special poverty-stricken communities have the congenital vulnerability with weak ability of preventing disaster and reducing disaster. On the contrary, just because of the insufficiency of feasible ability and its high vulnerability, it is easier to incur natural disasters. Once these natural disasters happen, they will bring great damages. Thirdly, the possibility of suffering from natural disasters and the occurrence of poverty in clustering contiguous special poverty-stricken communities are interacted. Because the poverty-stricken communities have the higher ecological vulnerability, therefore they are easier to be damaged by natural disasters; meanwhile, the communities with a high rate of disaster occurrence are always poor, they are often the poverty-stricken communities with bad natural environment. Judging from the associated analysis of disasters and poverty in clustering contiguous special poverty-stricken communities, disaster risk and poverty have a higher consistency in poverty-stricken communities. Fourthly, there are relative alternating property and potential cyclicity between the disaster-risk, the vulnerability, the feasible ability, impoverished, which is exactly reflected in clustering contiguous special poverty-stricken communities. On the one hand, when the natural disasters happen in the poverty-stricken communities, they will inevitably cause the enhancement of vulnerability in the poverty-stricken communities, and then they cause the decrease of feasible ability of reducing disaster, the drop of the feasible ability of rebuilding, since the external aid can't enter in the disaster-stricken areas, then the poverty-stricken communities become more impoverished. On the other hand, if we regard the "natural disaster-strong vulnerability, low feasible ability-poverty" as a vicious circle, when a new round of disasters happen, if we don't adopt the effective measures of disaster prevention and reduction as well as the risk management measures, it is easy to discover that, the poverty-stricken communities which have already fallen into poverty because of disasters will be reduced into a new round of natural disaster, and that will enhance their vulnerability and weaken feasible ability and the poverty will deepen continually, which will repeat in such circle. Fifthly, the continuation and the circulation of disaster and poverty in clustering contiguous special poverty-stricken communities, are mainly caused by the insufficiency of effective disaster risk management, and this also demonstrates the importance and necessity of strengthening the disaster risk management in clustering contiguous special poverty-stricken communities. In terms of path choice, we should strengthen the weather forecast, disaster warning and we should enhance and promote the capacity and technique of preventing disasters in the concentrated and contiguous special poverty-stricken communities. The most effective way is to further advance the construction of new countryside communities, to strengthen the feasible ability of poverty-stricken communities and reduce their vulnerability.

References:

[1] Ullrich Beck. Risk Society[M]. Translated and edited by He Bowen. Nanjing: Yilin Press, 2004.

[2] Ren Ze, Yu Zhen(ed.). Amartya K. Sen. Development as Freedom[M]. Beijing: Renmin University of China Press, 2002.

[3] Huang Chengwei. The Reconstruction Process and Challenges of Impoverished Villages after the Wenchuan Earthquake[M]. Beijing: Social Sciences Academic Press, 2011.

[4] Feng Quanguang. The Economic Synergistic Development at the Edges of Administrative Areas in Mountainous Region from the Geo-economic Regional Perspective: Taking the Wuling Mountainous Regions and Chongqing, Guizhou, Hubei and Hunan as Examples[J]. Journal of Mountain Science, 2009.

[5] Bai Jinxiang. Study on Economic Development of Mountain Fastness[M]. Beijing: The Ethnic Publishing House, 2006.

[6] Quan Fang. The Thoughts on Solving the Poverty of Multi-national Areas in Wuling Mountainous Region [J]. Journal of Guizhou Provincial Committee Party's School of CPC, 2008.

(Translated by Wang Dan)

Paradigm Shift of Disaster Risk Management: Looking upon Disaster Mitigation from the Perspective of Poverty Reduction*

Lv Fang

Abstract: Since the end of 20th century, there has been a distinct paradigm shift in the field of disaster risk management research. Early disasters are considered as the result of purely natural forces, so the responsibility of national disaster relief is more emphasized in the field of disaster risk management. With the development of science and technology, scientific paradigms have occupied the absolute dominant position in the field of disaster risk management. While the recent researches discover there is a high degree of correlation between disaster risk and the vulnerability of poverty population and poor communities. Thus, reducing the vulnerability of poverty population and poor communities becomes one of the important ways of disaster risk management. In this essay, disaster reduction programmes are proposed, which are aimed at poor communities and combined disaster

---

* This study is financed by the national post-doctoral Science Fund project—The practice performance assessment and policy suggestions of the policies to promote the whole villages in special types of poor areas (Project number: 20100480918) and the basic research project of the central colleges—The research on policy supporting system of avoiding calamity agriculture industry development in special types of poor areas (Project number: 120002040317). Lv Fang is a lecturer at Social Development and Policy Research Center of Huazhong Normal University.

risk management with sustainable development capacity construction of poor communities together. Considering this theory paradigm shift, there is no doubt that it will shed important enlightenment on the poverty alleviation work of concentrated special poverty-stricken areas in this new period.

Key words: Disaster Risk Management; Paradigm Shift; Vulnerability; Poverty Reduction

The well-known writer Ye Shuxian once put forward a view that "disaster and salvation" is one of the motifs of almost all civilizations in the world, when he gave a report in Sichuan University during the aftershocks of 2008. This view has inspirational significance, in an intuitive sense. There are two sides of meanings. On the one hand, the evolution of human society is almost always accompanied by the intrusion of disasters, which has been very clear in the era of origin of many civilizations; on the other hand, the cognition regarding to disasters will influence management strategy against disaster risks①. For

---

① It will be controversial that "the gods salvation" is regarded as a disaster risk management strategy. Perhaps, the research about Azande "witchcraft accused" in the North Africa by British anthropologists Evans Pulitzer, Chad (e. Evans-Pritchard, 1937) can explain it. A barn collapse crushed a person in the tribe by accident. It is only a simple thing, but Azande is sure that it is due to witchcraft. In the further study, Chad found, "witchcraft accused" actually refered to the specific social relationship structure, including the in-laws, the enemy and so on. If "witchcraft accused" is established, the next to do will be very clear. The clear point eased people the fear and the feeling of having no idea about what to do because of "emergency" and effectively maintained the unity of the tribe. Collective consciousness can be maintained and people also have some economic interests directly in punishing the witchcraft. For this, the idea of the gods salvation has the disaster risk management significance. Disaster events disorder the normal social order, whose influence is often beyond the control of human. The reasons for the disasters need a specific explanation, so as to maintain the exact knowledge and expectations for the future life "what to do". Like this, although China has the real disaster risk management "drought politics" since the Spring and Autumn Period, "the telepathy between heaven and the man" of Dong Zhongshu still has far-reaching influence. Once a disaster happens, emperors would need to discipline themselves and even worship, so as to defend the common ideas and the social order of the traditional society.

a long time in human history, disasters are viewed as "God's will", "God's action", "the events that no one can take responsibility for" ①. But this does not mean that practices of disaster risk management do not exist. For example, Chinese has attempted to construct water conservancy projects to protect agricultural production from the frequent floods and droughts as early as during the Warring States period. While at the national level, a relatively complete system of famine relief has been established to practice relief responsibilities of states. At the community level, through inter-relations such as family and geography, disaster response capacity has been strengthened to provide disaster relief. In the western world, national disaster relief and the religion organizations have played an active role for a very long time.

1. The Perspective of Scientism on the Disaster Risk Management Research

The term of "paradigm" comes from Thomas Kuhn, a philosopher of science, which presents certain value systems, conceptual tools and interpret paths which social science researchers are armed with. In early researches on disaster risk management system, the perspective of scientism occupied a dominant position, which is a kind of modern epistemology, advocating that the natural science and technology should be viewed as the foundation of the whole philosophy and it is believed that it can solve all problems.

In the field of natural science, disaster researches are to explore the factors that lead to disasters and affect the survival of human society. From the perspective of taxonomy, these factors are classified as external factors and anthropogenic factors. The former include floods, droughts, hails, debris flows, typhoons, earthquakes and tsunamis. The latter are caused by human activities, such as unrest, war, nuclear

---

① Li Yongxiang. *The Anthropological Research Review on Disaster*. Ethnic Studies, 2010, Vol, 3.

accidents and so on①. Besides, there are other classification methods, for example, according to the environment in which disaster-causing factors are formed, there are some factors in the atmosphere and hydrosphere such as typhoons, torrential rains, storm surges, tsunamis, floods and so on; there are some other factors due to the lithosphere activities as well as the biosphere accidents such as earthquakes, volcanoes, landslides, debris flows, diseases, pests and so on. These researches share a common purpose that it's possible to carry out purposeful disaster risk management through the mastery of awareness of dangers, the probability of occurrence, operational rules of the disaster-causing factors. In recent years, with the development of atmospheric research and the application of satellite technology, remote sensing simulation technology, radar technology, computer technology and so on in the field of scientific research on disaster, the cognition about disaster is updating and progressing. Obviously, there is one way of thinking the sense of the problem and solutions are both in the category of the scientism, which holds that human society can solve these problems through the scientific development through the understanding of these natural phenomena and mastery of its laws (even though the solutions may bring new problems, then there must be newer and better solutions). In the early 1990s, simple studies on disaster-causing factors faced great difficulties, the researchers found that studies cannot be separated from local, regional, and even more temporal and spacial factors to understand disaster-causing reasons, then they began to embark on the study of systematic environment.

Since the end of the last century, a large number of studies have shown that global warming issue has a significant impact on disaster occurrence mechanism. Scientists found that the world's warmest 10

---

① Despite the fact that there are some concerns about human factors which lead to disasters, but the screening of the human factors are mainly from the "macroscopic" level, which belongs to a kind of structural, or even superstructure factor.

years appeared after 1990 and showed an upward trend. Global warming and the change of climate and surface cover directly resulted in a period of frequent occurrence of the disasters. Global warming is attributed to disasters occurrence in many parts of the world. Although no one surely knows all influence that global warming will bring in, but "even according to the most conservative estimate, the forecast result will be surprising. Some forests will be increasingly endangered for some new and adverse growth conditions. Some cities will experience increasingly serious water shortages and heat wave attacks. The increase of temperatures and the change of rainfall pattern will add to the incidence of diseases, because the new conditions are more suitable for the reproduction of mosquitoes, ticks, rodents, bacteria and viruses. "The frequency of devastating storm will increase, large tracts of lowland coastal areas (a large number of people in the world living here) would suffer a serious flood threat when storm surges, or even be flooded. It is not hard to imagine that the situation that a whole country will be washed away by the flood will occur, such as the lowland Pacific island State of Tuvalu and Kiribati."①

Up to now, it is still difficult to comprehensively grasp the inner mechanisms of global climate change, but the research about the area under the influence of environmental factors and climate change has accumulated a lot. These studies focused on the specific natural landscapes and regional environment, through the medium and long term scale to recognize the relevance between climate change and the disaster occurrence to demonstrate the correlation between the two, and on the basis of these researches we can make adaptive adjustment to the relevant economic sectors. For example, owing to the continuing effects of drought, Iowa, the currently most major principal maize producing area of United States, will have to turn to plant wheat and cold-

---

① Charles Bullard. *Study Seeks Cut in Carbon Dioxide*. Des Moines Register, April 28, 1997, PP. 1M, 5M.

tolerance maize varieties, even if this led to the decline in unit production①. Correspondingly, the boundaries of disaster-causing factors discussion also enlarged to internal relevance perspective of the regional system, to explore "regional environmental stability and the spatial and temporal distribution of natural disasters; critical range evaluation of environmental changes caused by natural disasters; the reconstruction of the similarity of the distribution pattern of natural disaster in characteristic period (cold and warm periods, dry and wet periods), with its practical purpose to provide a basis for regional development planning for disaster reduction. " ②

It must be admitted that the system of scientific research on disasters is an important foundation for effective disaster risk management, with the development of scientific research on disasters, disaster response capacity of human society has greatly improved. It shows not only on the construction of large-scale infrastructure projects, but also on the control and intervention of certain types of disasters, such as artificial rainfall, artificial hail disasters and so on. However, the recent study found that disasters' appearance, or research on disaster-causing factors, cannot be separated from human activities. In other words, disasters have been caused by human activities. Theoretical paradigm of disaster risk management is gradually introduced into the reflection on human activities.

2. Vulnerability Analysis and Paradigm Shift of Disaster Risk Management

Today, vulnerability analysis is a classic method in many applied researches. It has a wide range of applications in finance, natural environment, computer science, marketing systems and other fields.

---

① Michael Bell. *An Invitation from Environmental Sociology*. Beijing:Peking University Press, 2010,P. 10.

② Shi Peijun. *Rediscuss on Theory and Practice of Disaster Study*. Journal of Natural Disasters, 1999, Vol, 4.

However, as for the vulnerability study theory and its originator, the term vulnerability really derives from social science research on disaster risk. In 1983, the world-renowned expert on disaster reduction, Frederick Kearney, in his masterpiece *Disasters and Development*, raised the issue of vulnerability. Kearney's inspiration came from the investigation about two earthquakes which had the similar magnitude but very different effects: In 1971, the city of San Francisco in California, United States, with 7 million population, encountered an earthquake of 6.4 on the Richter Scale. 58 people were killed in the entire process; two years later, a 6.2-magnitude earthquake occurred in Managua, the capital of the Republic of Nicaragua, which has slightly smaller intensity, but more than 6,000 people died in this earthquake. What causes the different disaster consequences under roughly the same disasters? During the questioning of disaster sensitivity, Kearney expounded his views on vulnerability theory. Kearney's research was quickly applied by international relief organizations, to improve the efficiency of their rescue and reduce future aid requirements[1]. Intergovernmental Panel on Climate Change (IPCC) soon applied vulnerability analysis in the field of climate change, and published its first assessment report in 1990, with preliminary expression about vulnerability of climate change. In assessment reports of 1996, 2001 and 2007, there was more complete understanding on vulnerability of climate change and attracted worldwide attention.

What should be noted is that vulnerability analysis is from the comparative study of different groups, "sensitivity of loss when facing certain risks or combat". It is not a series of specific interpretation of the events, and is mostly a basic structure of the analysis. It has a wide range of applications in different fields, such as analysis of vulnerability on specific ecosystem and certain groups. This structure of analysis not

---

[1] Martha G. Roberts & Yang Guoan. *The International Progress of Sustainable Development Research: A Comparison of Vulnerability Analysis and the Sustainable Livelihoods Approach*. Progress in Geography, 2003, Vol, 1.

only pays attention to the external factors leading to disasters, but also investigates the ability of groups to resist disasters, as well as the ability to recover from disasters. Their basic explanation can be summed up as: Risk=Hazard × Vulnerability①, which means that the extent of the disaster risk is equal to the results of potential hazards times vulnerability. On this basis, disaster risk management should be carried out by means of reducing vulnerability, strengthening response and recovery capacities of the target group, to achieve the goal of disaster reduction. In this respect, vulnerability analysis promoted the first theory paradigm shift in the field of disaster risk management, which is from concerns of external variable disaster-causing factors towards vulnerability analysis of overall structures. However, this paradigm shift of theory is relatively limited, although there are some concerns of the impact of human activities on ecological environment vulnerability, but the focus of interpretation continues to be the natural environmental system itself.

By the end of 20th century, the social science research on vulnerability had gathered to be a strong force, which promoted the paradigm shift on disaster risk management again. An important aspect of this change is the research on vulnerability overlaps. As far as the intention, vulnerability at first concerns the different groups whose social scientific significance are obvious. A medical anthropologist Professor in Tsinghua University, Jing Jun, who once put forward "Titanic laws" in the study on AIDS, thought that risk implementation and risk consciousness both showed a strong social stratum/class difference. Social groups that are at the bottom of society are also subjected to the invasion of risk②. From the perspective of the relationship between human activity and the environment, disaster

---

① Ben Wisner. "Vulnerability" in Disaster Theory and Practice: From Soup to Taxonomy, then to Analysis and finally Tool. International Work-Conference Disaster Studies of Wageningen University and Research Centre 29/30 June, 2001.

② Jing Jun. *The Titanic Rule: A Risk Analysis of the AIDS in China*. Sociological Studies, 2006, Vol, 5.

anthropology studies suggest that "disaster occurs as a result of combining environmental vulnerability with human population vulnerability, namely potentially damaging factors in ecological system and the conditions of the population in relatively backward state of socio-economic are its produce condition"①. The discussion of Baker on "Risk Society" is very inspiring, risks in modern society are more manmade risks which means that risk is systematically posed. Many disasters derived from human activities in the modern thinking mode of objective nature, and nature is the source of continually seizure for human to meet what they want to obtain. For a long time, nature keeps silence, but one day it carries out the massive retaliation. The marginalized and poor groups are more vulnerable when they face disasters, and the activities of poor groups, to a certain extent, increase the vulnerability of the natural environment.

As what has been said before, soon after vulnerability analysis appeared, it started to be widely applied to many research fields, such as ecological systems, climate change, agricultural development and poverty, but these researches are limited to their respective territories and rarely crossed. While a great number of studies have found that specific regions and specific populations are often influenced by a variety of potential risks, and because of human activities, vulnerability represents cause and effect practice form in the structure of the interaction among certain people, economic activities, social life, ecosystems and the natural environment. In this structure, human beings, as the most dynamic factor, become the key to understanding the problem, and also become the basic focus when seeking to improve the programs.

3. The Theory and Practice of Disaster Risk Management and Poverty Reduction

As Dr. Yodmani said, research paradigm about poverty and

---

① Li Yongxiang. *The Anthropological Research Review on Disasters*. Ethnic Studies, 2010, Vol, 3.

disaster risk management research changed at the same time in the 20 years after 1980. On the one hand, in a long period of time, disasters are regarded as contingencies which are caused by natural forces so the governments and disaster relief institutions must take some measures to deal with them. However, social and economic factors are rarely considered to be the causes of disasters. Yet with the further understanding of analysis of the structure vulnerability and the generating mechanism of vulnerability, the key of the disaster risk management is focused on improving living circumstances of disadvantaged people and marginal groups, so as to make them maintain their own sustainable survival and development and reduce the vulnerability of the environment. On the other hand, poverty research was measured by income standard in the early days, but many researchers and practical workers on poverty have found that it's necessary to build multivariate measure systems to understand the poverty problems on the basis of the reflection in the long term practice. Income standard is only the appearance of poverty, but the reasons leading to poverty are very complicated. Therefore, building multivariate poverty concept is needed. Multiplivariate concept explains poverty from many angles including politics, economy, culture and natural environment, but many poverty phenomena have close connection with disasters. Disaster is one of the important factors that lead to poverty, but the activities of the poor dramatically add to the environmental system's vulnerability. Therefore, the theory of combining disaster risk management with poverty reduction soon became a consensus in the related research areas and the international community. In 2005, the international conference of poverty reduction was held in Japan and passed the *Hyogo Declaration* which pointed out "confirm the inner link of disaster mitigation, sustainable development, poverty reduction and so on. And confirm the significance of absorbing all the related parties to participate, which includes the governments, regional organizations, international organizations, financial institutions, non-governmental organizations, the private sectors and the scientific communities". The conference adopted the outline for disaster mitigation action concerning the future 10 years: "The Hyogo Framework for Action 2005-2015: Building the Resilience of Nations and Communities against Disaster." which considered that "the increasing of loss due to disasters, caused serious consequence to the individuals, survival, dignity and livelihood and hard-earned achievement, especially to the poor". "The disaster reduction must be systematically incorporated into sustainable development and poverty reduction policies, plans and programs. At the same time it should be supported by bilateral, regional, international and partnership cooperation."

The announcement of *The Hyogo Declaration* and "The Hyogo Framework for Action 2005-2015: Building the Resilience of Nations and Communities against Disaster", marked the international community's theoretical consciousness about the combination of disaster risk management with poverty reduction practice, and it also called on countries to strengthen the poor people's ability against disaster and the ability for recovery, to defend human's survival, dignity and livelihoods. In terms of practice, there are many reference theory and experience about combing the disaster risk management with poverty reduction.

First of all, strengthen a nation's ability of disaster prevention and mitigation. Research on a nation's ability widely appeared in "development-oriented nation" discussions in recent years, but rarely emerged in disaster risk management. Generally speaking, a nation's ability includes the country's resource mobilization ability, organization and leadership skills and system construction ability. Dealing with disasters and reducing poverty efficiently reflects a nation's responsibility and its practice depends on sustainable capacity building. This ability includes affirming the correlation of disaster risk management and poverty reduction from strategic planning level of the country, formulating corresponding special regulations and systems in order to promote the implementation of sustainable development and the project which combines poverty and disaster risk, encourages related scientific researches and technological developmental work, invests more in the organization, management and resources.

Second, strengthen the integration of the disaster prevention and mitigation with poverty reduction, and develop department cooperation. Disaster risk management and poverty reduction are all complex affairs, but modern social administration and social labor division order carry out specialized management in the investment of the resources and the manpower. Thus it will become one of the important practices by establishing cooperation mechanism to integrate resources and promote the disaster risk management and poverty reduction progress. In recent years, non-governmental organizations worked more and have exerted good effect on the disaster rescue, the disaster prevention and mitigation, poverty reduction, thus forming an open platform. And the integration of positive energy is one of the important paths to promote disaster prevention and mitigation.

Finally, reduce the loss through the individual and the community ability construction. The individual and the community are closely related to the disaster risk, and they are also the ultimate parts of poverty alleviation. Enhancing the individual's ability of maintaining sustainable livelihood is helpful for them to get rid of poverty and enhance the ability to deal with the disaster risk and also reduce the influence of the poor people's inappropriate economic activities towards nature. On the other hand, only by the power from the state and the individual it is difficult to achieve the disaster risk management. The community education, community participation, community cohesion are also important because they can promote the supply capacity of public products to deal with emergency and disaster. Respect, protect and convert local knowledge, and make the community and the poor's key role reflected.

4. The Thinking of the Combination of Poverty Alleviation with Disaster Prevention and Mitigation in China's New Stage

China has made significant achievements in the poverty reduction.

Paradigm Shift of Disaster Risk Management: Looking upon Disaster Mitigation from the Perspective of Poverty Reduction

The world poverty population has actually reduced by 58 million in the past 25 years, while 70% of the achievement of poverty alleviation comes from China. As the biggest developing country, China has fulfiled the poverty reduction goal required by the *United Nations Millennium Development Goals* ahead of time. According to the measurements of the National Statistics Bureau and Chinese national poverty line, Chinese rural poor population has reduced from 250 million in 1978 to 14.79 million in 2007. The poverty rate has reduced from 30.70% in 1978 to 1.60% in 2007. Even according to international absolute poverty line (refers to the consumption is less than $1 per person a day), it was up to 730 million people in 1981 in China, but it reduced to 106 million in 2005, down with a decrease of 624 million. Poverty rate had reduced from 73.50% to 8.10% by the end of 2005. Though we made great achievements in poverty reduction, the task of poverty reduction is still arduous in the new stage. A prominent problem is that poverty population and ecological fragile areas, distribution highly presents "geographical coupling", according to the statistics of the National Environmental Protection Department in 2005, "95% of the absolute poverty population of the nation live in the areas which have very fragile ecological environment". These areas are more sensitive to climate change. Thus, coping with the impact of the disaster to poverty population will be an important task in the future work of poverty alleviation.

China is one of the countries which have most serious natural disasters in the world. Along with the global climate change, the fast economic, development, the speeding up of urbanization process, the stress on resources and ecological environment is intensified, and natural disasters prevention becomes more complex in China. Great importance has been attached to disaster prevention and mitigation work since the 1980s. Our country has promulgated more than 30 laws and regulations concerning disaster prevention and mitigation. All of these has promoted the disaster prevention and mitigation work into the legal system track. In March 1994, the Chinese government issued *The 21st*

Agenda of China, which defined the relationship of the disaster mitigation and ecological environment protection from national level. It's important content includes improving the management level against natural disaster, strengthening the system construction of disaster prevention and mitigation and reducing artificial factors that cause and aggravate natural disasters. At the beginning of the new century, guided by the Scientific Outlook on Development, both the national development strategy and local development planning pay more attention to the harmony between man and nature. In the past twenty years, China has made significant achievements in disaster reduction legal mechanism construction and integrated disaster mitigation capacity construction, initially forming a modern disaster risk management system. In the new period of time, the disaster prevention and reduction task in fragile ecological areas has become an important part of the disaster prevention and mitigation work in national level, which is not only related to life and property safety of local people, but also vital for whole country's ecological safety. As what was mentioned above, poverty population distribution is very dense in these areas. Through improving the poverty population's ability against disaster, not only the survival security can be maintained, but also the sustainable development and ecological protection can be promoted.

In a word, the national poverty alleviation work and disaster prevention and mitigation work can't be understood separately any more in practice in the new period. The ecological protection of the ecological fragile areas, ecological recovery tasks and poverty alleviation and development task should be combined to develop in practice. At the same time, we should realize that the shift of the disaster risk management paradigm "regard poverty reduction as disaster reduction", and the practice of the combination of poverty reduction with disaster reduction should be put in the specific situation of the poor areas and ecological fragile areas in China. Then it's necessary to understand the particularity of the area and the local population, and further form a practical road that is suitable for Chinese characteristics. This research is still not able to comprehensively solve this problem, but just to provide some ideas.

First, build strategic consciousness which combines poverty alleviation with disaster prevention and mitigation work. Honestly, in terms of dealing with the complex reality under the background of "geographical coupling" which means the combination of ecological vulnerability and the poverty problems, we still have no special mature experience or mode. In the past, poverty alleviation work and disaster prevention and mitigation work are developed independently. Poverty alleviation work focused on the poor people's income growth, while disaster prevention and mitigation work focused on the maintaining and recovery of the ecological environment. Generally speaking, the cooperation of the two aspects is not enough. The geography coupling characteristics in the poor areas with ecological vulnerability makes it very necessary to combine poverty alleviation with disaster prevention and mitigation, carry out the comprehensive management, develop the poverty population's sustainable livelihood, to realize the economic benefit, social benefit and ecological benefit together.

Second, correct previous doctrinal mode of development. In the mode, local affairs focuses on GDP, and ecological value, cultural value, social value are often marginalized. Since the 1990, China has deeply reflected on the doctrinal thinking about development, and proposed the Scientific Outlook on Development, seeking for inclusive economic growth. However, both local government and poor population are urgent to get in poor areas. Therefore, in the structural industrial transfer background, they are easily restricted to developmental doctrine thinking, thus return to the old path to sacrifice the ecological environment in exchange for temporary economic growth. Therefore, we should avoid the short-term development thinking in terms of regional development planning.

Third, the differentiation principle of policy support. The poor and ecological fragile areas are mostly inhabited by ethnic minorities, accompanied by the complexity of natural and geographical conditions and the diversity of its economies and social cultures. So, we should not make the developmental policy of the region with "one-mode-for-all". The policy support with differentiation is based on local resources, tries to tap native potentials, integrat local resources to find a development road which is suitable for the local characteristics.

Finally, reflect the initiative of ethnic minority areas, poverty population, communities and the ecological environment. To some extent, on the previous developmental path, the minority areas, poverty-stricken population, poor communities, the ecological environment are all in a passive position "voiceless" about the development which leads to a lot of problems we are facing now. However, in the new period, the poverty alleviation and disaster prevention and mitigation work are just focusing on these groups which are traditional marginalized and objectified. How the traditional knowledge of ethnic minorities, poverty-stricken population, communities and the ecological environment devote themselves to the process of development, and the benefits they deserve must be carefully taken into consideration.

References:

[1] Shi Peijun. Rediscuss on Theory and Practice of Disaster Study [J]. Journal of Natural Disasters, 1996 (4).

[2] Jing Jun. The Titanic Rule: A Risk Analysis of the AIDS in China [J]. Sociological Studies, 2006(5).

[3] Ye Shuxian. Disaster and Salvation in Literature[J]. Ziguangge Magazine, 2008(18).

[4] Li Yongxiang. The Anthropological Research Review on Disaster [J]. Ethnic Studies, 2010(3).

[5] Li Yongxiang. An Anthropological Study on Mud-flow Disaster: A Case of Ailao Mountain of Yunnan[J]. Ethnic Studies, 2008(24).

[6] Michael Bell. An Invitation from Environmental Sociology[M]. Beijing: Peking University Press, 2010.

[7] Ben Wisner. "Vulnerability" in Disaster Theory and Practice in From Soup to Taxonomy, then to Analysis and Finally Tool[R]. International Work-Conference

Disaster Studies of Wageningen University and Research Centre, 29/30 June,2001.

[8] Martha. G. Roberts & Yang Guo'an. The International Progress of the Sustainable Development Research Method: The Comparison of Vulnerability Analysis Method and Sustainable Livelihood Method[J]. Progress in Geography, 2003 (1).

[9] David Alexander. Disaster Management: From Theory to Implementation [J]. JSEE: Spring and Summer. 2007,1(9).

[10] Kasperson, J. X & R. E. Kasperson. Global Environmental Risk[M]. New York: United Nations University Press, 2011.

[11] Evans Pritchard, E. E. Witchcraft, Oracles and Magic among the Azande [M]. Oxford: The Clarendon Press, 1937.

[12] EL-Sabh M. I. & T. S. Murty. Natural and Man Made Hazards[M]. Dordrecht,Holland:D. Reidel Publishing Company, 1988.

[13] Suvit Yodmani. Disaster Risk Management and Vulnerability Reduction: Protecting the Poor[C]. The Asia and Pacific Forum on Poverty: Reforming Policies and Institutions for Poverty Reduction Held at the Asian Development Bank, Manila, February 5-9, 2001.

(Translated by Zhao Huihui & Kong Xiang)

The Unconventional Action on Impoverished Villagers' Housing Reconstruction in Earthquake-stricken Areas

—Taking Makou Village in Sichuan Province as an Example

Chen Wenchao[*]

Abstract: In the post-disaster reconstruction process, the villagers of the impoverished villages in the earthquake-stricken areas impoverished villages know that reconstruction can lead to poverty or can make them return to poverty. But the state intervention, the political social mobilization and the institutional arrangement construct the legitimacy of reconstructive actions and mould the reconstructive field. The reconstructive subjects invest all their strength into the housing reconstruction project, and have generated a kind of unconventional action supported by the social capital operation which makes the impoverished villages' post-disaster reconstruction in the earthquake-stricken longlasting areas become possible. However, when the effects of the practices are concerned, the exterior nature of the non-conventional action have to be further noticed. The rise in price of market building materials and the lack of the longlasting effective

---

[*] Chen Wenchao is a doctoral student at Sociological Academy of Huazhong Normal University and his main research focuses on sociological theories and its application.

mechanism of the action makes the non-conventional characteristics more prominent in the collective non-conventional action, which plays an important role in maintaining persistent and stable post-disaster reconstruction.

Key words: Earthquake-stricken Areas; Villagers of Impoverished villages; Housing Construction; The Non-Conventional Action

1. How the Non-conventional Action Becomes Possible

Because of the high magnitude and the strong destructive force, the Wenchuan earthquake resulted in the collapse of more than 17 million rural houses including 2.673 million peasant houses, 0.92 million livestock homes and 1.178 million injured households in the post-disaster reconstruction planning areas[1]. Therefore, in the process of the post-disaster reconstruction, housing reconstruction not only becomes the primary target of the reconstruction, but also becomes the main social action of the peasants in reconstructing area in reality. From the social action perspective, the logic of this kind of housing reconstruction action—the non-prospect lending reconstruction and poor peasants racing to reconstruct their houses in spite of the increasingly rising price of building materials—is non-logical[2], according to the definition about farmers' actions and characteristics by Chayanov school and Adams Smith, Marx, Engels' petty-farmer school, and the representations of the occurrence and development of the peasant action process by Schultz's petty-farmers school[3]. In the concrete process of the interaction between individual and society, action and structure, the

---

[1] The Development-Oriented Poverty Alleviation Leading Group of the State Council & The Poverty-Stricken Village Post-Disaster Reconstruction Planning Working Group: Overall Plan for Wenchuan Post-Earthquake Poverty-Stricken Village restoration and reconstruction, 2008.

[2] Pareto. V. *Sociology Writings*. London: Pall Mall,1966,P. 300.

[3] Huang Zongzhi. *Peasant Family and Rural Development in the Yangtze Delta*. Beijing: China Book Company,2000.

mandatory character of the earthquake disaster makes us realize the importance of a clear post-disaster reconstruction planning in our practice. For everyday life, survival is not only the lowest level of the demand, but the most basic form of the demand ①. Housing collapse endangers the villagers' survival. As the farmers said, "there is even no place to stand". Therefore, in order to ensure a place to live, housing reconstruction has become one of the necessary actions for villagers in the earthquake-hit areas. From the motivational perspective farmers' survival needs of the earthquake-hit area prompt the corresponding motivation to rebuild houses, and make the reconstruction action rational to a certain degree. This inference will not lead to the generalized "non-logical behavior" but a normal rational choice. So, if we just qualify it from this perspective we cannot deny that is a kind of instrumentally rational choice②. But if we take the farmers' economic capacities and daily life background into account, the fact is just the opposite—housing reconstruction over one's capacity is unwise.

Rational choice theory obviously loses its discourse function in explaining this action. In the Western theory, this paradoxical social fact is consistent with the aberrant behaviors elaborated by Durkheim and Merton, and the irrational behaviors elaborated by Weber and Habermas. We can also call it non-routine action relative to "routine action" and "habitus" in the modern sociological language ③. But this non-routine action is not the "social anomie" in the course of the practice function. On the contrary, just under this non-conventional action's influence, the rise of the housing reconstruction just inversed the logic of market economy, and makes our post-disaster reconstruction project run smoothly. Therefore, to some extent, non-conventional action makes the implementation of housing reconstruction after the disaster in

---

① Pareto. V. *Sociology Writings*. London: Pall Mall, 1966, P. 215.
② Max Weber. *Economy and Society*. Beijing: China Commerce and Trade Press, 1998, P. 56.
③ Zhang Zhaoshu. *Non-routine Action and Social Change: A New Concept and Topic*. Sociological Studies, 2008, Vol, 3.

the quake-stricken areas become possible. Analyzing from the action unit of the peasant and looking into the particular social structures and situation, we can find that in the process of the interaction of the action and the structure, individual and society①, the non-conventional action just makes up for the defects and inadequacy of the conventional action or the rational action, and shapes our daily life system. But in the reconstruction of Wenchuan earthquake-stricken areas, how the non-conventional action becomes possible and how it achieves the "Pareto Optimality" showes significant internal tension, this seemingly paradoxical topic inspires the thought of conduct analysis and discussion in depth in order to seek the source of the non-conventional action, and establish new rules of action system and operation mechanism, further to increase the social progress as well as reducing social costs.

In order to deeply think of the non-conventional action, and then to build the panoramic framework, this paper aims to explore the formation mechanism of the non-conventional action. To this end, we take the "social science starts from the practice road" as the foothold and starting point②, analyze the reconstruction process about Makou Village in Sichuan Province, concentrate on the description and analysis of the generating path of the non-conventional action, so as to develop the "concrete" cognitive about it, and construct the new motivation mechanism, realize the leap-style development in earthquake-hit villages so that the harmony and development can be maintained in post-disaster impoverished villages.

2. Social Mobilization: Legitimacy Construction of the Non-conventional Action for Housing Reconstruction

In the process of the construction and development of harmonious

---

① Zheng Hangsheng & Yang Min. *The Construction and Development of Sociological Theory System : An analysis of the Meaning of the Relationship between Person and Society in the Research of Sociological Theory*. Sociological Studies, 2004, Vol, 2.

② Huang Zongzhi, *Understanding China : Social Science that Starts from Practice*. Social Sciences in China, 2005, Vol, 1.

society, post-disaster reconstruction has become the social action that affects peasants' well-being from the national level. A strong government led by a strong party conduct national relief and the reconstruction project which is combined with poverty alleviation and development by strong social mobilization capability. " Not allow one person to starve to death, not allow one person to freeze to death" has not only become this campaign's bottom line, but also the party and the state's minimum requirements of the social risk control. For the impoverished peasants in the earthquake-hit areas, this kind of discourse which can promote peasants' social welfare is always controlled by the policy executors, and what they can do is "look at the rice pot". If they have no rice, like that "one can not make bricks without a straw", they can do nothing. From the statistical data of Makou village, the phenomenon of causing and backing into poverty by the earthquake increases greatly: the impoverished population increased from 15 households before the earthquake to 87 households① after the earthquake; per capita net income decreased from ¥ 2,450 before the earthquake to ¥ 1,580 after the earthquake, per capita decreased by ¥ 870; the grain yield per capita was 450kg ②. Therefore, because of the limited source of income and the unclear prospects, although peasants' strong housing demands can only be castles in the air. So, in particularly impoverished spatio-temporal domain, if we want to make housing reconstruction possible and prompt impoverished peasants to actively involve in housing reconstruction project, we must stress the

---

① During the earthquake and the aftershock, in Makou village, as houses of brick structure and reinforced concrete structure are suffered small effect, most of them have cracks only, they belong to the type of repair and reinforcement; houses of civil engineering structure are the main source of dilapidated houses, and usually poor population live in such kind of house. According to peasants' logic: With enough money, we will build a house early, and with no money, we only live in the adobe house.

② The Development-oriented Poverty Alleviation Leading Group of the State Council & The Poverty-stricken Village Post-Disaster Reconstruction Planning Working Group: Overall Plan for Wenchuan Post-Earthquake Poverty-Stricken Village Restoration and Reconstruction, 2008.

reconstruction of the mainstream discourse through the social mobilization strategies to form the housing rebuilding legitimacy.

Historical experience shows that the identity of the authority and the state power gives and maintains the high trust towards the government. In the process of post-disaster reconstruction, the party and the government's social mobilization therefore have great charismatic power. Although it differs from the "error recognition" phenomenon ①, the impoverished peasants of earthquake-stricken areas recognize the force as the corresponding grace but not a kind of power. As the Makou villagers said, "what the government says is certainly right, so we just follow it". Indeed, in order to better promote the post-disaster reconstruction work, all functional departments—from the country, province, city and township to village—are conducting propaganda the significance of the post-disaster reconstruction and relevant policies in various effective forms. In terms of the practical effects, this approach enhances the peasants' confidence of reconstruction, and further more encourages public participation.

In Makou village, peasants can hear and understand the official reconstruction policies through the morning broadcasting, and can also know related reconstruction projects through government officials and non-governmental organization workers. Although this kind of social mobilization can satisfy peasants' reconstruction interests and demand, and effectively inspire the reconstruction passion, enable them to better participate in the reconstruction process, these have been taken into account by government social policy makers, and not just for this aspect. If we analyze this kind of social mobilization in light of the action unit of villagers, then it has another meaning. According to the theory of the hierarchy of needs, we can clearly find that other levels of demand are impossible when needs for survival is threatened ②. In the

---

① Bourdieu & Hua Kangde. *The Practice and Reflection: Guide of Reflection of Sociological Theory*. Beijing: Central Compilation & Translation Press, 1998, P. 186.

② Maslow. *The Motivation and Personality*. Beijing: Huaxia Press, 1987, P. 60.

concrete operating process, individual social action can only incline to aspects related to survival, and others cannot be taken into account at all. From this point of view, we can find that in the current spatio-temporal domain, peasants' main goal also focuses on the housing reconstruction, especially for those peasants whose houses were dilapidated, it seems more pressing. In short, peasants will get down to the housing reconstruction even without certain social mobilization and publicity. So, the key point of the significance of this large-scale social mobilization is to give some legal significance to housing reconstruction. From the previous analysis, we know that most of the households that fell and backed into poverty by the earthquake have been attributed to absolute poverty households. They have no savings, just can afford the food and clothing, and can't get much money through going out for a job[1]. If there are children in school, then this will make the family even more difficult to get rid of the economic pressure and fall into the survival dilemma for a long time. Therefore, in consideration of the social resources deficiency in the impoverished people's everyday life, the meaning endowed by the state policy of social mobilization lies in a certain legitimacy of the reconstruction actions; especially during the economic difficulties the reconstruction meets the social and personal requirements, and the needs of poor villagers' interests. In other words, the non-conventional action of the housing reconstruction is a legitimate action advocated by the party and government, but not an irrational behavior. For those dilapidated households whose survival have been threatened, although this kind of authority cannot completely make them get rid of poverty and become better off, it is at least the

---

[1] According to the statistical data of Makou village, the whole village has 210 households, and the population is 767; it has 450 labor force and 250 peasant-workers who are outside of their home years, which accountes for 56% of total labor force. They are generally coal mining and construction workers, and their per capita area of farmland is 1.22 mu. So, every household has 3.65 people on average, and two of them are labor force, and every household has a peasant-worker work outside annually to maintain the family livelihood, and this is one of Makou village's characters.

life-saving straw.

Although not by the way of "trying to persuade sentimentally and rationally", the whole social mobilization process is a legitimate construction. In Makou village, the thought of "recall and learn from painful experience, don't wait and rely on external help, but actively self-help against earthquake" makes the legitimacy of housing rebuilding deeply engraved in the villagers' hearts. Especially, for the three kinds of people who are unable to carry out reconstruction, such as immigrant workers, disabled people and old people who have no relatives to rely on, cadres take the initiative to contact with them, analyze policies together, so that the villagers can share social policy welfare to the greatest degree. The government actively encourages the households who are able to rebuild their houses. If the households have labors working outside, the government tries its best to contact them, and persuades them to return to hometown and join in the reconstruction; for the households which are difficult to realize high standard reconstruction, the government relaxes the policy to allow them to reconstruct a house; for those who have no ability to reconstruction, such as the old, the government takes the special measures such as centralizing them to some safe places. During the whole mobilization process, ideological work is the main way of legitimacy construction. In addition, the construction of the legitimacy also cooperate with certain material guarantee and support, otherwise the social mobilization and the legitimacy construction will be questioned by public. In Makou Village, the government made corresponding commitment especially about the payment of the reconstruction cost. The promise to the village is to help the local economic development, such as breeding pigs etc. In the course of the investigation, we found that some villagers doubted the legitimacy and quitted the housing reconstruction. From this special case, we found that in particular poor spatio-temporal domain, rebuilding legitimacy is not only a social policy welfare sermon, but more as a legitimacy reconstruction based on the farmers' consideration about the reconstruction. If the construction succeeds, peasants will fully believe in the rebuilding action, and certainly be able to actively participate in the reconstruction project, which shows that the non-conventional action can be put into practice. Conversely, if the reconstruction fails, this would run against villagers' desire, and restrain villagers' actions, the housing reconstruction can only be a castle in the air. Therefore, in order to ensure the construction of the legitimacy, the government also made corresponding systematic arrangements in the implementation process.

3. Reversed Construction Schedule: The Institutional Arrangement of the Non-conventional Action

As the intermediary of thinking and action, discourse reflects the thinking and guides the action. When this kind of discourse mobilization rises to the institutional level by the authority and power, it becomes a kind of arrangement. In accordance with Hayek's words (the master of liberalism) "a series of institutional formations which have clear objectives, is extremely complex but ordered. However, this is not a result of the design, not a result of the invention, but a result of human actions from those who are not explicitly aware of the outcome"[1]. From the institutional perspective, this kind of arrangement has a certain rigidity, this means that, if we take actions in accordance with the arrangements, then the system will meet the needs of the individual, and the social order will keep fine; but if not in accordance with the arrangement, the power utilization will certainly deprive individuals' corresponding right. Especially under the arrangement of the "reversed construction schedule", this kind of constraint will be stronger.

In the earthquake-stricken areas, the reconstruction has become a whole nation event, and is the livelihood issue that guarantees the disaster victims can share the achievements of the social development. *Report on the Work of the Government* in 2009 pointed out one of the

---

[1] F. A. Hayek. *The Constitution of Liberty*. Beijing: SDX Joint Publishing Company, 1997, P. 67.

The Unconventional Action on Impoverished Villagers' Housing Reconstruction in Earthquake-stricken Areas

primary tasks is to basically complet the reconstruction of those houses which were collapsed or seriously damaged because of disasters so as to ensure that the affected people can live in new premises by the end of this year①. In the bureaucracy management mode, the dominant thought of the grass-roots government will be more clear, and the officials must remain committed to the higher authorities' tasks and targets, or their own future will not be guaranteed. With the basic unit administrative personnel's words, "the reconstruction is linked with our reward, and if we can not complete the task ahead of the deadline, our reward and promotion will be affected; it's compulsory rule, with various standards placed there. In the concrete implementation process, we can only use the subsidy standards to achieve the task." "For those households who want to give up this kind of subsidized housing, we leave no stone unturned to mobilize them, and the branch secretary of village party and the township government staffs take turns to communicate with them. If it doesn't work, they are forced to take actions." Their words reflect the functions of the "reversed construction schedule" in non-conventional action. That's why the reversed construction schedule makes the non-conventional action become inevitable.

In the concrete practice process, the village's reconstruction arrangement is detailed to every tiny step. This kind of arrangement provides that the housing reconstruction is not only related to the damage degree, but also related to the people of a household. From the investigation, villagers reflect that subsidies for the household with three or less members is ¥ 16,000, with four to six people is ¥ 19,000, and with more than six people is ¥ 23,000②. In this systematic arrangement, social structure correspondingly becomes the power

---

① Wen Jiabao. Report on the Work of the Government—Delivered at the Fifth Session of the Eleventh National People's Congress on March 5,2009.

② Similar to this kind of arrangement, in Makou Village, the subsidy standards of households with houses to be repaired also take this way to carry on, and there are three grades: ¥1,200, ¥2,500, and ¥4,000.

source of the action, and the "timely assistance" to the dilapidated households stimulates the peasants' social action. Rebuilding homes with the help of about ¥20,000 can at least guarantee a tiny bit of place to live in, and also means meeting the minimum level of survival needs. However, the social fact is quite ideal, in the case of equity bewteen the rights and obligations, the power must be in concert with the duty, or the rights will be taken back. Similarly, while villagers accept the reconstruction subsidy money, they obey the corresponding requirements of housing reconstruction operation such as observing the institutional requirements, or the rigidity level of it will not be reflected, otherwise the rigidity of the system will restrict the action. The reconstruction subsidy is not a one-time payment to households, but is given to the households by two times, and each amount is different, mainly depending on the standard of housing construction progress. When a household was identified as reconstruction households, the willing to rebuild their houses does not mean that they can receive the national housing reconstruction aid; only when they have completely rebuilt the foundation of houses can they receive a corresponding 80% of the funds. If they can't get sufficient funds to continue to rebuild their houses, then the remaining 20% of the funds will be unable to obtain. Only when the main project has been built can the remaining funds be completely given out. In short, under the lead of people's livelihood, orderly solving the basic problems closely related to the disaster area peasants' mass lives has become the main content of the institutional arrangement, and only when the institution implementation condition suits well, the interest can be fully vested in the people.

　　In a sense, reversed construction schedule makes the post-disaster reconstruction within a certain time and space possible. Just imagine, there are still a number of households who are able to build houses at the absence of "inverted construction deadline", both the absolute poor and the relatively poor households have certain ability to conduct such social action. As without the limit of time, villagers can first save enough money to lay the foundations of houses, and after building the foundations, they will receive 80 percent subsidies from the government to build the body of the houses. It also makes the "habit" of ideas in our daily life even become a kind of Habitus. But for reconstruction as a whole, this is also the largest disadvantage. After all, we are faced with poorer people in quake-hit areas. Therefore, it also proves such a truth in the fields of social science research that only by directly entering a certain situation and observing the action that occurs, can we truly understand the meaning of these actions[1]. And just under the institutional arrangements, this kind of non-routine action has a clear manifestation, that is the appearance of inverted construction deadline. The so-called inverted construction deadline mainly refers to the progress of a project which is calculated by the end date. In the process of post-disaster reconstruction, inverted construction deadline becomes the most extensive means in the reconstruction of many areas. In the Makou village, housing reconstruction project also must be carried out in accordance with the progress of institutional arrangements, or the possible subsidies will also vanish like soap bubbles. As farmers said, if the government requires that reconstruction should be completed by certain time, then we must get it done ahead.

In the theory of practice, Bourdieu regards "field" as a network, a structure that is continuously constructed. Further more, he believes that every "field" is a unique space, a unique circle, and is also a game with its own different rules. In his view: Field is not a dead structure, not an empty place, but is a game space, those individuals who believe and pursue reward participate in this game [2]. For peasants in housing reconstruction, the "inverted construction deadline" under the institutional arrangements is also a specific field of the game. Only by following the appropriate rules, can they obtain the corresponding

---

[1] Deng Jin. *The Interpretative Communication of Activism*. Chongqing: Chongqing University Press, 2004, P. 91.

[2] Bourdieu & Hua Kangde. *The Practice and Reflection: Guide of Reflection of Sociological Theory*. Beijing: Central Compilation & Translation Press, 1998, P. 132.

benefits. If this effect is enlarged, we can see that "inverted construction deadline" with evidence makes the post-disaster reconstruction possible. However, in the specific practice, the institutional arrangements of "inverted construction deadline" is not the only condition, without social mobilization, reconstruction would be like a tree without roots, water without a source. In particular without specific social support, it is like a castle in the air. After all, the decision of social action lies in the hands of the majority of social action.

4. Patchwork: Capital Support of Non-routine Action in Housing Reconstruction

Institutional arrangement can prompt poor villagers in quake-hit areas to take non-routine action in housing reconstruction, but this does not mean that the system limitations and restrictions will lead to unconventional action. After all, it also depends on the farmer's own conditions. Especially for poor regions, the emergence of unconventional actions requires a certain amount of material conditions support, or even if the system is rigid and strong, no matter how the social mobilization operate in any case, it is still far away from reality, or is out of reach. In one word, capital support plays a decisive role. If capital is sufficient, then people will naturally carry out housing reconstruction, otherwise the reconstruction is still difficult. In the survey, we often hear such a statement that "the boss of brick factory will not give you bricks free of charge, you have to buy something in cash". and so on. This may show that the existence of the market makes the currency in circulation inevitable, which must rely on certain currency symbols, otherwise people will find it difficult to buy building materials for housing reconstruction. Therefore, judging from the reconstruction costs, lack of start-up capital is an important bottleneck for public housing reconstruction.

The support of capital comes firstly from the low interest loans granted by government.

On the whole, low-interest credit loan is the largest preferential policy for the households in the process of post-disaster reconstruction. In accordance with appropriate standards, one household of reconstruction can borrow or loan ￥30,000, and the loan can only be used in housing reconstruction. This low-interest loan is relatively easy to get in terms of procedures. The villagers just need to ask the Village Committee to write a letter as an introduction, and then to the Township Government to perform the appropriate procedures, and finally to County Savings Bank. Compared to the previous loans, low-interest loans by government is not only in low interest rates, but the procedure is easier and more convenient in the course of our investigation, farmers also have similar reflections. In terms of interest, the loan interest under the institutional arrangement is half cut, the interest now is 5.4‰ instead of 18 ‰ before, calculated by this interest, a quarterly loan of ￥ 30,000 now is ￥ 648①.

With the help of the national housing reconstruction grants and low-interest loans, the poor people in quake-hit areas are still lack of start-up capital for reconstruction. Neither government subsidy nor credit loan, can fully solve the problem of capital insufficiency. Judging from the cost of a house, in accordance with the standards of the new rural construction②, the price of a rough house without decorating is 100,000 yuan, even with government subsidy of 20,000 yuan, and credit loan of 30,000 yuan, but the total 50,000 yuan and that is only half of the whole cost, and the remaining 50,000 yuan becomes the plight of farmers. In the Makou village, farmers' field played an

---

① But under the government-dominated loans, it does not mean that we have to hold a mortgage. In the real life, most of the villagers in the Makou village had mortgage with their new houses. This also means that if we can't pay off loans in a certain time, we must have mortgage loan with our own houses.

② Overall plan on Wenchuan Post-earthquake Restoration and Reconstruction for Poor Village fomulates that the combination of post-disaster reconstruction with poverty alleviation and development is the clear demands of the Party Central Committee and the State Council, and it is the pressing needs of people in poor areas and the inevitable trend of China's economic and social development.

important role in the process of resolving this dilemma. ①

Farmers clearly know where they need to spend money, and where they need to invest in the economic calculation for housing reconstruction. In the Makou village, farmers have a more clearly economic idea of their own. For them, this thinking is also mainly from the idea of mutual assistance, that is, relying on social networks to gather sufficient resources. For the individuals in Chinese society, the approaches to get resources or capital are mainly from their own social networks such as geographical, occupational, blood relations and so on②.

For poor villagers in the earthquake-stricken areas, geographical and occupational relations may not work. As villagers live in the same poor area, and suffer from earthquake disaster. All need to rebuild the damaged houses. From this perspective, geographical network is impossible to work, let alone the social network such as occupational relations. Just imagine, for farmers in the same poor village, living by planting crops, this means that capital stock is limited, and they themselves can't produce "reciprocal" remaining resources. Even the migrant workers that have established a certain social network outside the village, but due to the strengthening of modern social mobility, this social network is full of risk and uncertainty, so it is difficult to obtain appropriate resources from the mobile social network. Therefore, from the perspective of indigenous knowledge, the best access to appropriate and greater resources can only rely on blood relationship. In the process of investigation toward the Makou village, the distribution of resources was mostly through blood relationship channels. In the primary group, blood relationship has two channels, on the one hand, it can be

---

① In Bourdieu's view, the field refers to the spatio-temporal location of a person, which can bring people some certain capital. Now, according to special conditions in village, we use farmer's social field as we think that the family in village consist of relationship between man and women, and they both have their own subject status and social capital.

② Zhai Xuewei. *The Logic of the Chinese Action*. Social Sciences Academic Press, 2002, P. 144.

expanded through the male side, such as his brothers and sisters; on the other hand, it can be spread out through females' relationship, such as wife's mother, sisters and so on. For example, in the survey, a villager told us, "my house now has cost about 65,000 yuan including 30,000 yuan from bank loans. As I got loans earlier, in July last year, when the countries had not yet started this policy, so the interest was still as high as 18%, and I had paid 1,630 yuan for the interest before the new year. Now the state has adopted this policy, interest rates has reduced to 5%, which means that we can pay less. I borrowed the remaining money from my wife's relatives, her brother and sister's situation was slightly better than us for the earthquake damage they suffered was not serious, so we borrowed each of them 10,000 yuan." Through these words, we can clearly see that building house requires certain capital support, and without start-up funds, it is still difficult to get the subsidy from institutional arranged grant money, which leads to "Matthew Effect" in the process of rebuilding. Therefore, using one's own social capital to borrow money is becoming a more beneficial and efficient way.

Compared with the loan, the biggest benefit of this patchwork of borrowing is that villagers don't need to return some interest; and as for repayment, there are no certain guarantees or corresponding time limits. But this does not mean that villagers do not need to repay money, if they don't repay money, it will produce a more adverse impact. In other words, borrowers do not need to return money within a certain time, they can have more time to pay back. According to Chinese action logic, if borrowers don't have money at hand, they can return later, and when lenders are in need of money, borrowers must do their best to help them. In the Makou village, one villager told us, "Relatives played a crucial role when I was in trouble, and I would return money as they also earned money uneasily. I borrowed a little as their situation was better, but they were still poor. Because of the relationship of relatives, and our poor situation, they had done their best to help us. So we must pay back every cent we borrowed, let alone they didn't require interest. Otherwise, it is hard for us to be relatives later again. And the cost of the behaviour of disrepayment is too much to borrow money from anyone next time."

Non-routine action becomes possible in the earthquake-hit areas due to the capital support of this action, otherwise it is just like the castle in the air. With the full support of the capital, villagers can generate more secure reconstruction of unconventional action. But the reason why we call it non-routine action is that we must explain it based on real-life background. Using the method of expanding individual case, we can find in the context of the global financial crisis, the shrinking of working economy, as well as the tightening of export industries, the economic outlook was uncertain, people in poor areas who take the initiative to carry a certain amount of debts must naturally need a lot of courage and ability. This emphasizes unconventional characteristic of non-routine action that it doesn't follow the routine and common sense.

5. Generation of Non-routine Action of Building Houses and Practical Outcome

With the help of policy mobilization, inverted construction deadline and effective capital, the unconventional action of poor villagers become possible in housing reconstruction of quake-hit areas, and also develop an effective mechanism of action, either in national reconstruction planning pilot village like Makou village, or in the non-planned villages, as long as there are policy mobilization strategies, accompanied by efficient capital support, the intersection among the three parts is the necessary and sufficient condition of unconventional actions and only when the three parts reach an effective balance, can therefore post-disaster reconstruction be realized, as shown in the figure below:

In the Makou village, according to statistics, there were 253 villagers who work outside the village especially those who have technical expertise returning home during the post-disaster reconstruction, and they joined the reconstruction of new homes actively. Among them, the number of project managers, steel workers,

The Unconventional Action on Impoverished Villagers' Housing Reconstruction in Earthquake-stricken Areas

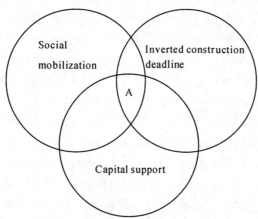

Figure Formation mechanism of non-routine action of building house①

welders, electricians, silicon engineers, carpenters, plasterer and other technical personnel was 52, and who were "indigenous experts "or "new force " for reconstruction. On the choice of construction mode of rural houses, Makou village chose to rebuild permanent houses for the whole 93 households, which adopted the approach of "Unified planning, joint households construction, unified construction and allocation". In the implementation, they adopted an approach of "five unification, three open, one household". "Five unification" refers to unify plan and building design, level the ground thogether, unify materials procurement, unify organization of construction, and unify quality control. "Three open" consists of three parts, and the first is to open the post-disaster reconstruction policy. The second is to open post-disaster reconstruction projects and programs. The third is to open public material procurement and financial management of funds. "One household" means raising funds by each household itself, then they submit these money to the financial group for rural housing reconstruction, the financial group will publicize the funds raising and

---

① A indicates non-routine action, its location is the intersection of social mobilization, institutional arrangements and capital support. This indicates that social mobilization has strong power and wide range, but due to the limited power of the institutional arrangements and capital support, the non-routine action occurs only in the more balanced state of the three parts.

using to the masses. In the first stage of reconstruction, 93 households in the Makou village need to rebuild permanent houses, and the first floor of the main project of 28 suite permanent rural houses in centralized settlements have been completed. The second layer is going to be covered with tiles, and the remaining 65 households are also actively preparing for leveling the foundation. There are about 84 households which have finished houses' maintenance and reinforcement among the 117 households. When we entered the village to start the investigation, centralized settlements housing construction had been completed, and many households had moved into the residence. Judging from this scene, non-routine action played a greater role in the housing reconstruction and produced the desired results.

As we see effects of non-routine action in the process of post-disaster reconstruction, we also have to take into account the unexpected adverse effects of this unconventional action at the same time. In the course of the investigation, we clearly feel that non-routine action contributed to the post-disaster reconstruction, but since this non-routine action has formed a kind of social fact, even within a certain time and space it has turned into a collective reconstruction action. Based on the inherent requirements of the law of the market, we can clearly find the inherent contradiction between supply and demand, and the intensification of contradictions lead to the rising prices in market. In this regard, while the countries have taken some administrative measures for regulation, it is still difficult to contain interests drive, especially for the price of construction materials, as shown in the table.

With the increase of market price of construction materials and housing costs soaring, it correspondingly adds the burden of the reconstruction subject. As government subsidies haven't changed, price subsidies can only fill the space of the increase. Even for the general maintenance, subsidies are still difficult to meet the cost of maintenance. However, in view of the real life, even in this situation of the market economy, there are still many people in the queue to purchase

Table Building material market price in the Makou Village in Sichuan province (unit: yuan) ①

	Unit	Price before the earthquake	Price after the earthquake	Rise(%)
Red brick	Bolt	0.38	0.58	52.6
Concrete iron	Ton	105	165	57.1
Cement	Ton	340	380	11.8
Stone	Vehicle	180	360	100.0
Red tile	Piece	1.0	1.75	75.0
Floor	Piece	105	140	33.3
Wood	Cubic meter	600	800	33.3
Skilled worker	Day	50	80	60.0
Unskilled labourer	Day	30	50	66.7

materials for building houses. To a certain extent, it also reinforced our knowledge about the non-routine action of these subjects in the process of reconstruction in the earthquake areas. In other words, supported by the social mobilization, institutional arrangements and social capital, this non-routine action becomes one of the most important types of action in the process of reconstruction after the disaster. Reflecting on the reconstruction process, we are more aware that housing reconstruction of non-routine action must be combined with the long-term mechanism. Once villagers lack certain degree of support, these actions will certainly result in loss of confidence and recognition as they don't repay money in time, and thus threaten the reconstruction and order of production, and threat social stability and construction of civilized harmonious communities.

(Translated by Li Ping)

---

① Price here includes freight price, and freight price is not the same as the distance is different, but for the same village, freight price is roughly the same and there is no big difference.

References:

[1] Chen Wenchao. From the Perspective of Sociology of the Peasants' Consumption of Present Conditions and Characteristics[J]. The World of Survey and Research, 2005(1).

[2] Chen Wenchao. The Means: Survival Logical of the Social Vulnerable Members [J]. Journal of Yunan University of Nationalities, 2008(1).

[3] Deng Jin. The Interpretative Communication of Activism[M]. Chongqing: Chongqing University Press, 2004.

[4] Development-oriented Poverty Alleviation Leading Group Office of the State Council & Poverty-stricken Villages Post-Disaster Reconstruction Working Group. Overall plan for Wenchuan Poverty-Stricken Village Post-earthquake Restoration and Reconstruction. 2008.

[5] Guo Yuhua. "The weapon of the weak" and "Hide text"[J]. Reading, 2002 (7).

[6] Huang Zongzhi. Peasant Family and Rural Development in the Yangtze Delta [M]. Beijing: Zhonghua Book Company, 2000.

[7] Huang Zongzhi. Understanding China: Social Science That Starts from Practice[J]. Social Sciences in China, 2005(1).

[8] Development-oriented Poverty Alleviation Leading Group Office of the State Council. Disaster Reconstruction Planning at the Village Level of Makou village, LiZhou District, Sichuan province (2008-2010), 2008.

[9] Zhai Xuewei. The Logic of the Chinese Action[J]. Beijing: Social Sciences Academic Press, 2002.

[10] Zheng Hangsheng & Yang Min. The Construction and Development of Sociological Theory System: An Analysis of the Meaning of the Relationship between Person and Society in the Research of Sociological Theory[J]. Sociological Studies, 2004 (2).

[11] Zheng Hangsheng. Zheng Hangsheng's Academic Course of Sociology [M]. Beijing: The Renmin University of China Press, 2005.

[12] Zheng Hangsheng. People's Livelihood Should be Put in a More Prominent Position in the Post-disaster Reconstruction [J]. Truth Seeking, 2008 (15).

[13] Zhang Zhaoshu. Non-routine Action and Social Change: A New Concept and Topic [J]. Sociological Studies, 2008 (3).

[14] Wen Jiabao. The Government Work Report-March 5, 2009 at the 11th National People's Congress of the Second Meeting. 2009.

[15] Anthony Giddens. Composition of the Society [M]. Translated by Li Kang and Li Meng. Beijing: SDX Joint Publishing Company, 1998.

[16] Bourdieu & Hua Kangde. The Practice and Reflection: Guide of Reflection of Sociological Theory [M]. Beijing: Central Compilation & Translation Press, 1998.

[17] Anthony Giddens. Nation: State and Violence [M]. Beijing: SDX Joint Publishing Company, 1998.

[18] Max Weber. Economy and Society [M]. Beijing: China Commercial & Trade Press, 1998.

[19] Max Weber. The Methodology of Social Sciences [M]. Beijing: Huaxia Press, 1999.

[20] Maslow. The Motivation and Personality [M]. Beijing: Huaxia Press, 1987.

[21] Parsons. The Structure of Social Action [M]. Translated by Zhang Mingde, Xia Yunan and Kang Peng. Nanjing: Translation Press, 2003.

[22] Bourdieu. The Sense of Practice [M]. Translated by Jiang Zihua. Nanjing: Translation Press, 2003.

[23] Hayek. The Constitution of Liberty [M]. Beijing: SDX Joint Publishing Company, 1997.

[24] Alexander L. Action and Its Environments [M]. New York: Columbia University Press, 1988.

[25] Giddens. Modernity and Self Identity: Self and Society in the Late Modern Age [M]. Cambridge: Polity Press, 1990.

[26] Harbermas. The Philosophical Discourse of Modernity [M]. Boston: MIT Press, 1988.

[27] Pareto V. Sociology Writings [M]. London: Pall Mall, 1966.

(Translated by Li Ping & Chang Xiaochong)

Postscript

In recent years, China has experienced Wenchuan earthquake, Yushu earthquake, droughts and other massive natural disasters, people's life and property, means of livelihood, social life and the psychological health suffered major blows in these disasters. Practice shows that disaster and poverty have a high correlation, and natural disaster is one of the main reasons for the rural poverty and the return to poverty. How to manage the risk of disaster is a big challenge that needs to respond under the background of global climate change in the strategy of China poverty reduction and development, even in the whole public governance system. At the same time, in China's great practice of disaster relief, reconstruction after disaster and poverty alleviation and development, the uniqueness and superiority of Chinese cultural tradition, political system, social mobilization mechanism and economic development mode has also attracted extensive attention from international society, and become a kind of valuable resources which need to be exploited in the field of global disaster risk management and poverty reduction.

International Symposium on Theories and Practices of Disaster Risk Management and Poverty Reduction, was co-hosted by International Poverty Reduction Center in China and The United Nations Development Programme (UNDP) on April 14 and April 15, 2011.

It has exchanged the experience of poor village reconstruction after Wenchuan earthquake, and based on it the participants has widely and deeply discussed on disaster risk management and poverty reduction theory, practice and policy framework. It is hoped that the seminar will promote the combination of the disaster risk management and poverty reduction and development strategy, planning and policy system, enhance the poor communities' capability toward disaster prevention and development. This book is a collection of the papers and materials for exchange in this symposium.

The Chief Editor sorted and sifted the whole conference papers and materials for exchange, and modified the style and language of selected articles in respect to the author's original meaning. The authors are responsible for points of view of every article, which do not represent the host and sponsor of the conference. The authors and translators of the articles are tagged at the beginning or the end of each article. Dr. Li Xiaomei and Dr. Guo Jin are responsible for English translation check.

This book is the second book of *Series of Theory and Method on Disaster Prevention and Mitigation, Post-disaster Reconstruction and Poverty-reduction and Development* sponsored by International Poverty Reduction Center in China and The United Nations Development Programme (UNDP). The first collection *Reconstruction of Impoverished Villages after Wenchuan Earthquake: Progress and Challenges* was published in January 2011 by Social Science Academic Press (China), which is also a collection of relevant conference achievements. Comparatively speaking, the first collected works focuses on summing up the reconstruction experience of impoverished village in Wenchuan earthquake stricken area, while the second focuses on analyzing and thinking based on the summarizing, and hopes to give extensive discussion on poverty alleviation and sustainable development in post-reconstruction era.

The book has the full support of Huazhong Normal University Press Co., Ltd. Ms. Feng Huiping, the executive editor, has devoted a lot of hard work to this book, and Wu Dan, a post graduate student, has assisted the chief editor a lot in editting and publishing the book. We wish to take this opportunity to extend our heartfelt thanks.

Huang Chengwei and Lu Hanwen
Dec. 2012